A U.S. Carbon Cycle Science Plan

Anna M. Michalak, Robert B. Jackson, Gregg Marland,
Christopher L. Sabine, and the Carbon Cycle Science Working Group[1]

August 2011

A report of the University Corporation for Atmospheric Research supported by NASA, DOE, USDA, USGS, NOAA, NSF, and NIST. The views expressed herein are those of the authors and do not necessarily represent the views of the agencies, or any of their sub-agencies.

---

[1] Robert F. Anderson, Deborah Bronk, Kenneth J. Davis, Ruth S. DeFries, A. Scott Denning, Lisa Dilling, Robert B. Jackson, Andy Jacobson, Steven Lohrenz, Gregg Marland, A. David McGuire, Galen A. McKinley, Anna M. Michalak, Charles Miller, Berrien Moore III, Dennis S. Ojima, Brian O'Neill, James T. Randerson, Steven W. Running, Christopher L. Sabine, Brent Sohngen, Pieter P. Tans, Peter E. Thornton, Steven C. Wofsy, and Ning Zeng.

A U.S. Carbon Cycle Science Plan was prepared by the University Corporation for Atmospheric Research under award number NA06OAR4310119 from the National Oceanic and Atmospheric Administration, U.S. Department of Commerce. The statements, findings, conclusions, and recommendations are those of the author(s) and do not necessarily reflect the views of the National Oceanic and Atmospheric Administration or the Department of Commerce.

This is PMEL contribution number 3731.

Table of Contents

Executive Summary

Understanding of the Earth's carbon cycle is an urgent societal need as well as a challenging intellectual problem. The impacts of human-caused changes on the global carbon cycle will be felt for hundreds to thousands of years. Direct observations of carbon stocks and flows, process-based understanding, data synthesis, and careful modeling are needed to determine how the carbon cycle is being modified, what the consequences are of these modifications, and how best to mitigate and adapt to changes in the carbon cycle and climate. The importance of the carbon cycle is accentuated by its complex interplay with other geochemical cycles (such as nitrogen and water), its critical role in economic and other human systems, and the global scale of its interactions.

The need for improved understanding of the global carbon cycle and better research coordination led to the development of the first U.S. Carbon Cycle Science Plan, published more than a decade ago. That document outlined a plan for land, atmosphere, and ocean observations; manipulative experiments; and Earth-system modeling to improve our understanding of the contemporary carbon cycle and our ability to predict its future.

The development of a new Plan was initiated by the U.S. Carbon Cycle Interagency Working Group (CCIWG) and the Carbon Cycle Science Steering Group (CCSSG), and outlines a strategy for refocusing U.S. carbon cycle research based on the current state of the science. The development of this Plan was led by a committee of 25 active members of the carbon cycle research community, and the result is intended to provide U.S. funding agencies with information on community-based research priorities for carbon cycle science over the next decade. The Plan emphasizes the long-lived, carbon-based greenhouse gases, carbon dioxide (CO_2) and methane (CH_4), and the major pools and fluxes of the global carbon cycle. The recommended research is global in scale, and there is therefore a strong need for international cooperation and collaboration.

While many of the research goals in the 1999 Science Plan remain important for the coming decade, new research thrusts are also needed. These thrusts include a more comprehensive look at the effects of humans on carbon cycling, including the consequences of carbon management activities; the direct impacts of CO_2 on ecosystems and their vulnerability or resilience to changes in carbon and climate; a quantitative understanding of the uncertainties associated with the carbon cycle; and the need to coordinate researchers from the natural and social sciences to address societal concerns.

The Plan is organized around three overarching questions:

Question 1. How do natural processes and human actions affect the carbon cycle on land, in the atmosphere, and in the oceans?

Question 2. How do policy and management decisions affect the levels of the primary carbon-containing gases, carbon dioxide and methane, in the atmosphere?

Question 3. How are ecosystems, species, and natural resources impacted by increasing greenhouse gas concentrations, the associated changes in climate, and by carbon management decisions?

In addition, the Plan recognizes the central role of sustained observations that underlie all of the outlined science objectives. There is need for an optimally designed and integrated system for long-term observations, data collection, and data management.

Incomplete representations of the carbon cycle cause large uncertainties in estimates of future changes in the climate system. Conversely, uncertainties about future climate also make it more difficult to predict future changes in the carbon cycle. In balancing the global carbon cycle and gaining a process-level understanding of its components, it is important to evaluate, understand, and deal with the uncertainty that arises through measurements, models, analyses, and projections, and the complex interdependence of the carbon, climate, and socioeconomic systems.

The overriding science questions provide basic long-term direction for guiding carbon cycle research. To make progress toward answering the questions, and to provide guidance for continuing research, we have outlined six science goals that should be pursued over the next decade. These six goals (together with references to the overriding questions they are primarily designed to address), are:

> **Goal 1 (Q1, Q2):** *Provide clear and timely explanation of past and current variations observed in atmospheric CO_2 and CH_4 – and the uncertainties surrounding them.*
>
> The scientific community needs to be able to provide the broader public with a clear and timely explanation of past and current variations observed in atmospheric CO_2 and CH_4, as well as the uncertainties surrounding these explanations. We note that 'timely' is an important part of this goal. To serve public policy needs, atmospheric observations and clear analyses are needed in close to real time. To address this goal, we need to develop the capability to accurately estimate variability in carbon sources and sinks as well as the processes controlling that variability.

> **Goal 2 (Q1, Q2):** *Understand and quantify the socioeconomic drivers of carbon emissions, and develop transparent methods to monitor and verify those emissions.* This goal seeks to derive process-level understanding of the human processes and motivations that determine carbon emissions from energy use, industrial activity,

and land use. Improved understanding will enable better evaluations of current emissions levels and better projections of future emissions, including the implications of alternative policy scenarios. Atmosphere-based measurements, remotely-sensed observations, evaluation of socioeconomic parameters, and other tools need to be developed to provide confirmation and confidence in mitigation commitments. The institutions and infrastructure for monitoring and verification of international agreements must come from the national and international political processes, but the tools and methods need to be developed by science.

> **Goal 3 (Q1, Q2, Q3):** *Determine and evaluate the vulnerability of carbon stocks and flows to future climate change and human activities, emphasizing potential positive feedbacks to sources or sinks that make climate stabilization more critical or more difficult.*
>
> All carbon reservoirs and carbon processes are not equally vulnerable to change, resilient to stress, responsive to management, or susceptible to unintended side effects of management decisions. We need to be able to identify which carbon pools and flows are most vulnerable and to understand the physical, chemical, and biological processes important in determining the degree of vulnerability of these pools and flows. We also need to predict the consequences of carbon management and sequestration schemes on vulnerable pools and to support carbon management goals by prioritizing the resources that are needed to assure the stability of the most vulnerable stocks and flows.

> **Goal 4 (Q3):** *Predict how ecosystems, biodiversity, and natural resources will change under different CO_2 and climate change scenarios.*
>
> The direct effects of elevated greenhouse gas levels, along with the accompanying changes in climate, are likely to alter ecosystems profoundly on land and in marine and freshwater environments. Beyond the interaction with climate change, there is a need to assess the direct impact of increasing atmospheric greenhouse gas concentrations on ecosystems, beyond their potential role as carbon reservoirs or sinks. Three examples of such impacts are altered marine ecosystem structure due to ocean acidification, biodiversity impacts on land and in the ocean, and the potential

stimulation of net primary productivity due to additional CO_2. The interacting effects of climate and biogeochemistry need to be understood.

Goal 5 (Q1, Q2, Q3): *Determine the likelihood of success and the potential for side effects of carbon management pathways that might be undertaken to achieve a low-carbon future.*

This goal is especially important as concerns increase over anthropogenic impacts on the atmospheric concentrations of greenhouse gases and their impacts on the global carbon cycle. There is a need to understand interlinked natural and managed systems sufficiently for individuals, corporations, and governments to make rational and well-informed decisions on how best to manage the global carbon cycle, and especially the anthropogenic impacts on this cycle.

Goal 6 (Q1, Q2, Q3): *Address decision maker needs for current and future carbon cycle information and provide data and projections that are relevant, credible, and legitimate for their decisions.*

The scientific community needs to provide carbon cycle information needed by decision makers and other stakeholders, understand how decision making affects the evolution of the carbon cycle, and determine how information about the carbon cycle can be relevant to policy decisions. Meeting the needs of decision makers requires an interactive process in order to understand those needs. This goal also recognizes the need to be anticipatory. The needs of decision makers a decade from now will not necessarily be the same as the needs they confront now and a goal of research is to anticipate and probe creatively so that we are prepared to confront tomorrow's questions.

A number of key cross-cutting research components comprise the central core for advancing carbon cycle science over the next decade, and these have been grouped into four high-priority elements. These elements embody the action items of carbon cycle research, with each of them contributing to all six research goals. The first element encompasses sustained and focused observations, which include atmospheric, ocean/coastal/inland water, terrestrial ecosystem, demographic/social, and remote-sensing observations. The second element

includes studies of system dynamics and function across scales, including intensive process studies and field campaigns, manipulative laboratory experiments, and manipulative field studies. This work should be designed as coordinated, integrative studies across traditional disciplinary boundaries where appropriate and possible. The third element focuses on modeling, prediction, and synthesis, including improving existing models, adding human dimensions to Earth system models, and augmenting synthesis activities. Finally, the fourth element centers on communication and dissemination, including improving dialogue among the decision-making community, general public, and scientific community, developing appropriate tools for communicating scientific knowledge to decision makers, and evaluating the impact of scientific uncertainty on decision making.

Interdisciplinary studies and improvements in both inclusion of, and collaboration with, the social and political sciences are essential to the success of this Plan. Visions of the future need to be strengthened through interactions with integrated assessment efforts and studies of carbon management. Similarly, the increasing importance of international collaboration is also apparent. U.S. scientists need to participate and take leadership roles in international assessments and syntheses, field campaigns, model inter-comparisons, and observational networks. Such international participation offers opportunities to leverage investments in resources and to contribute the knowledge and creativity of U.S. scientists to coordinated research.

The conduct of science depends on the institutions and structures that support the research. Institutional structures and opportunities to improve coordination and to ensure the achievement of the Plan's research goals include:

- Providing more opportunities for sustained, long-term funding.
- Enhancing carbon cycle data management.
- Encouraging directed calls for integrated topics in carbon cycle research, including research in the social sciences.
- Facilitating efforts to contribute to integrated, interdisciplinary efforts such as the assessments of the Intergovernmental Panel on Climate Change.

- Establishing stronger links between the CCIWG of the U.S. Global Change Research Program and other U.S. interagency working groups focused on climate change and mitigation.

- Developing a strong connection between carbon cycle research and the developing ocean acidification program.

- Expanding the North American Carbon Program to a new Northern Hemisphere Carbon Program.

- Improving international linkages.

- Using the North American Carbon Program and Ocean Carbon and Biogeochemistry program as models to initiate similar, problem-oriented research communities, including the creation of a group with strong roots in both the social and natural sciences.

- Implementing a process for periodic measurement and evaluation of progress in pursuing the goals of this Plan.

- Continuing to provide broad support for education and training, with an increased emphasis on interdisciplinary education focusing on carbon/climate science and decision making in a global context.

The overriding priority detailed in this research Plan is to develop and maintain a broadly-focused, balanced, integrated research agenda. Along with our emphasis on CO_2 and CH_4, additional non-greenhouse gases, such as carbon monoxide (CO) and the ratio of oxygen to nitrogen (O_2:N_2), provide important constraints on the global carbon cycle and are part of the plan in that context. Consideration of the greenhouse gas nitrous oxide (N_2O) and other non-carbon greenhouse gases is essential, but beyond the scope of this Plan. In general, connections between the global carbon cycle and the cycles of water, plant nutrients, and oxygen will need to be made to round out our understanding of the controls on the global carbon cycle, but these are not directly included under this Plan. Our intention is that complementary studies will be linked to the carbon cycle research proposed here to provide a broader understanding of the global carbon cycle and other biogeochemical cycles. Finally, throughout this document we emphasize the importance of an integrated system to collect and maintain the essential data that drive scientific understanding.

The Plan outlined here must be implemented efficiently and effectively. It is clear, however, that the breadth and intensity of the research agenda will depend on the resources available. We estimate that the total U.S. carbon cycle budget will need to be increased to approximately $500 million per year, not including platform costs (e.g., satellites, ship time, aircraft time), to achieve the goals outlined in this Plan. The interdependence of the many components of this research Plan is critical and the final approach needs to maintain balance among the various research foci, within the resources that are available. Greater commitment of resources will allow more complete understanding sooner, to the benefit of society as a whole. The importance of carbon cycle research within the pressures of confronting global change justifies this accelerated commitment of resources.

Chapter 1
Introduction

Carbon is an integral part of the Earth system and the building block of life. Its presence in the atmosphere as carbon dioxide (CO_2) and methane (CH_4) is critical to maintaining a habitable climate. It plays a major role in the chemistry, physics, and biology of the oceans. Carbon is currently also the dominant element in human energy use, forming the basis of coal, oil, and natural gas – hydrocarbon compounds derived from plants that removed CO_2 from the atmosphere hundreds of millions of years ago. The continual transport and transformation of carbon in the Earth's atmosphere, rivers and oceans, soils, rocks, living organisms, and human systems is what we call the global carbon cycle.

Human activities have substantially altered the Earth's natural carbon cycle. Human use of energy has grown exponentially in the last century, and the extraction and combustion of fossil fuels have replaced society's early reliance on renewable energy sources such as biomass, wind, and running water. As a result, CO_2 is building up in the atmosphere and in the oceans. In addition, land use for activities such as farming, forestry, and urbanization has gradually released soil and plant carbon, further increasing the amount of carbon in the atmosphere and oceans.

Overall, the mixing ratio or concentration of CO_2 in the atmosphere has increased to more than 390 parts per million (ppm) since the start of the industrial revolution. Today's value contrasts with the 280 ppm CO_2 or less in the atmosphere for at least the previous 800,000 years, and the concentration is now increasing by an average of almost 2 ppm per year (IPCC, 2007). Methane concentrations have increased proportionally more, to 1.8 ppm now from 0.8 ppm before the industrial era.

These changes in atmospheric concentrations of CO_2, CH_4, and other greenhouse gases are affecting the Earth in at least two important ways. First, our climate is changing. The Earth's surface is, on average, 0.8°C warmer than it was 100 years ago, and the most recent decade was the warmest in at least a century (Arndt et al., 2010). Other aspects of climate, such as the amount, distribution, and timing of rainfall, are also changing. The impacts of anthropogenic climate change are widespread (e.g., Jones et al., 2009). Second, increased atmospheric CO_2 has direct effects on both terrestrial and aquatic ecosystems. On land, increasing CO_2 can alter plant productivity and biodiversity as well as the competitive success of weeds and other species. In the oceans, the pH of ocean surface waters has already decreased by about 0.1 pH unit since the start of the industrial revolution (Bates, 2007). This acidification of the oceans imperils many calcifying marine organisms, including corals, shellfish, and marine plankton, which form their skeletons or shells out of calcium carbonate. This, in turn, is likely to have significant economic consequences for the fishing and tourism industries (Fabry et al., 2008).

Understanding the Earth's carbon cycle is therefore an urgent societal need as well as a challenging intellectual problem. The impacts of human-caused changes in the global carbon cycle will be felt for hundreds to thousands of years. Direct observations and process-based understanding are needed to determine how the carbon cycle is being modified, what the consequences of these modifications are, and how best to mitigate and adapt to changes in the carbon cycle and climate. The importance of understanding the carbon cycle is accentuated by its complex interplay with other important geochemical cycles, such as those of nitrogen and water, its critical role in economic and other human systems, and the global scale of its interactions.

The most widely known evidence for the changing global carbon cycle is the ongoing record of atmospheric CO_2 concentrations begun at Mauna Loa in Hawaii in the 1950s and continuing today at more than 100 sites around the world. These data, combined with economic data for fossil fuel use and historic data on land use, show that approximately half of the CO_2 emitted through fossil fuel burning and land use change has remained in the atmosphere. The other half has been removed by the Earth's oceans and terrestrial biosphere. The processes and motivations governing CO_2 emissions and uptake are not fully understood. As a result, our projections of the future carbon cycle are poorly constrained, yielding large uncertainties in the future trajectory of climate change (e.g., Friedlingstein et al., 2006). Our understanding of the magnitude and variability of emissions of CH_4 and its oxidation to CO_2, as well as the processes controlling their variability, lags even behind that of CO_2.

The need for better understanding and coordination of global carbon cycle research led to the first U.S. Carbon Cycle Science Plan, published about a decade ago (Sarmiento and Wofsy, 1999). That document outlined a plan for land, atmospheric, and ocean observations, manipulative experiments, and Earth-system modeling to improve our understanding of the contemporary carbon cycle and our ability to predict its future. The 1999 Science Plan focused primarily on quantifying the oceanic and terrestrial carbon sinks, with the goal of balancing the global carbon budget and quantifying the Northern Hemisphere carbon sink.

Although the carbon cycle research community has developed much improved global- and regional-scale carbon budgets over the last decade, a new U.S. Carbon Cycle Science Plan is now needed to address evolving science and policy priorities.

With greenhouse gas concentrations rising rapidly, active management of the global carbon cycle is increasingly urgent, but management without understanding can be ineffective or even counter-productive. Carbon cycle research is therefore needed on the efficacy and environmental consequences of carbon management policies, strategies, and technologies.

Humans are an integral part of the carbon cycle, both through their influences on relatively 'natural' and managed ecosystems and through direct emissions of greenhouse gases from energy and industrial systems. Study of the human elements of the carbon cycle needs to be more explicitly included in carbon cycle research.

Ecosystems, species, and natural resources are increasingly affected by rising greenhouse gas concentrations, climate change, and carbon management decisions. Research that focuses on important climate and ecosystem thresholds and tipping points (i.e., situations where a gradual change in an important control results in an abrupt change in a system) (e.g., Raupach and Canadell, 2007) is needed, as well as research to better understand the direct impacts of increased CO_2 and CH_4 concentrations on ecosystem form and function. An example of a potential climate threshold or tipping point is the possibility that the warming of permafrost soils and the accompanying runaway release of CO_2 and CH_4 to the atmosphere would lead to a cycle of additional warming.

Finally, decisions about the carbon cycle will inevitably be made with imperfect knowledge. We need a better understanding of uncertainty in the global carbon cycle and its environmental implications. We need ways to reduce uncertainty when appropriate and to deal with uncertainty when necessary.

The reassessment of U.S. carbon cycle science priorities described in this document was initiated by the U.S. Carbon Cycle Interagency Working Gr oup (CCIWG) and the Carbon Cycle Science Steering Group (CCSSG) in 2008 (see Appendix A for the charge to the committee). This new Plan is intended to provide U.S. funding agencies that support carbon cycle research with guidance on research priorities in carbon cycle science for the next decade. The research to be supported is global, and the need for international cooperation and collaboration is acute. A committee of 25 natural, physical, and social scientists was assembled to focus this effort (see Appendix B), using extensive outreach to garner input from a broad research community (see Appendix C).

The new Plan presented in this document describes research that is global in scope and collaborative in scale. Without a thoughtful Plan articulating research priorities, it would be difficult to coordinate activities among disciplines, researchers, and research sponsors, as well as provide appropriate resources for research. It is important to note that this document is not an implementation plan, and does not include the detail necessary to fully characterize specific research and resource needs. We anticipate that the implementation plans of existing programs will be revised to reflect the new priorities, and new implementation plans will be developed to identify project-level needs based on the recommendations presented here. Any mention of specific projects in the current document is intended for illustrative purposes only.

As a framework for the revised research agenda, Chapter 2 provides a brief history of the 1999 Science Plan, progress made since that plan was prepared, and the overall context in which our new Plan has been developed. In Chapter 3, we describe the overriding questions that guide the new research agenda. We then identify specific goals and achievable objectives for the next decade and beyond in Chapter 4 and outline some primary research elements that we believe must be pursued to achieve these goals in Chapter 5. These elements are the basic research components needed for developing the science. In Chapters 6 and 7, we turn to some challenges and opportunities for the success of the Plan. In Chapter 6 we characterize some of the collaborations and cooperation – both international and interdisciplinary – that are necessary for success. Finally, Chapter 7 summarizes the vision of the scope and priorities of the needed research.

The new Plan emphasizes the long-lived, carbon-based greenhouse gases CO_2 and CH_4, and the major pools and fluxes of the global carbon cycle. Certain non-greenhouse gases, including carbon monoxide (CO) and the ratios of oxygen to nitrogen ($O_2:N_2$), provide important constraints on the global carbon cycle, and are part of the Plan in that context only. Consideration of the greenhouse gas nitrous oxide (N_2O) and other non-carbon greenhouse gases is essential, but this Plan focuses specifically on carbon-based species. In general, expanded research on connections between the global carbon cycle and the cycles of water, nutrients, and oxygen will need to be made to enable comprehensive understanding of the controls on the global carbon cycle, but this work is also beyond the scope of the current document, and therefore not directly included under this Plan. Throughout this document, we emphasize the importance of an integrated system to collect and maintain the essential data that drive scientific understanding. Our hope is that complementary studies will be linked to the carbon cycle research proposed here to provide a broader understanding of the global carbon cycle and other biogeochemical cycles.

Chapter 2
History and Context

This chapter reviews the basic structure of the 1999 U.S. Carbon Cycle Science Plan, some aspects of the 1999 plan implementation, and key planning documents that have been published since the release of the 1999 plan. Together, these documents and associated activities provide the context for the development of the new Plan. This chapter is intended as background information, whereas discussion of the new Plan begins in Chapter 3.

2.1 The 1999 U.S. Carbon Cycle Science Plan

In 1998 and 1999, a working group of 16 carbon cycle researchers prepared a science plan to coordinate carbon cycle research in the United States. The intent was "to develop a strategic and optimal mix of essential components, which include sustained observations, modeling, and innovative process studies" (Sarmiento and Wofsy, 1999). In the ensuing decade, carbon cycle researchers have worked to improve carbon-observing networks and to coordinate research projects addressing the plan's goals. Considerable progress has been made, but constraints on funding and time have prevented some parts of the plan from being fully realized. Research over the last decade has also identified new issues that were not highlighted or foreseen in the 1999 plan. For instance, concerns about how the human perturbation of the carbon cycle might affect the overall distribution of carbon pools have intensified, and there is increasing public policy interest in options for mitigating human impacts and managing the carbon cycle.

The 1999 Science Plan posed two fundamental science questions: 1) What has happened to the CO_2 that has already been emitted by human activities (past anthropogenic

CO_2)? and 2) What will be the future atmospheric CO_2 concentration trajectory resulting from both past and future emissions? These questions focused on the past, present, and future atmospheric CO_2 concentrations. To address these questions, the 1999 plan articulated five program goals that guided the U.S. carbon cycle research program into the 2000s:

Goal 1: Quantify and understand the Northern Hemisphere terrestrial carbon sink.

Goal 2: Quantify and understand the uptake of anthropogenic CO_2 in the ocean.

Goal 3: Determine the impacts of past and current land use on the carbon budget.

Goal 4: Provide greatly improved projections of future atmospheric concentrations of CO_2.

Goal 5: Develop the scientific basis for societal decisions about management of CO_2 and the carbon cycle.

2.2 Implementation of the 1999 Science Plan

After the 1999 U.S. Carbon Cycle Science Plan was completed, it formed the basis for the carbon cycle chapter of the broader Strategic Plan for the Climate Change Science Program (Climate Change Science Program, 2003). In that document, the two overriding questions of the 1999 plan were updated to:

- How large and variable are the dynamic reservoirs and fluxes of carbon within the Earth system, and how might carbon cycling change be managed in future years, decades, and centuries?

- What are our options for managing carbon sources and sinks to achieve an appropriate balance of risk, cost, and benefit to society?

The North American Carbon Program (NACP) and the Ocean Carbon and Climate Change (OCCC) program were developed to address Goals 1 and 2, respectively, of the 1999 plan. The U.S. research agencies also made significant investments in additional projects aimed at the other three goals of the 1999 Science Plan.

The *North American Carbon Program Science Plan* (Wofsy and Harriss, 2002) and the *Science Implementation Strategy for the North American Carbon Program* (Denning et al., 2005) outlined "carbon cycle research focused on measuring and understanding sources and sinks of CO_2, CH_4, and CO in North America and adjacent oceans." The NACP is organized around four questions:

- What is the carbon balance of North America and adjacent oceans? What are the geographic patterns of fluxes of CO_2, CH_4, and CO? How is the balance changing over time? ("Diagnosis")
- What processes control the sources and sinks of CO_2, CH_4, and CO, and how do the controls change with time? ("Attribution/Process")
- Are there potential surprises (could sources increase or sinks disappear)? ("Prediction")
- How can we enhance and manage long-lived carbon sinks ('sequestration'), and provide resources to support decision makers? ("Decision support")

Research activities were recommended and prioritized within each major area to contribute to an integrated and well-tested system for observing, understanding, and predicting carbon fluxes over North America and adjacent ocean regions, and for providing timely and useful information to policymakers based on the results.

An integrated, multi-agency implementation strategy for oceanic observations and research was developed in parallel with the NACP to determine how much CO_2 is being taken up by the ocean at the present time and how climate change will affect the future behavior of the ocean carbon

sink. Within the broader goals outlined by the 1999 Science Plan, the document *Ocean, Carbon and Climate Change: An Implementation Strategy for U.S. Ocean Carbon Research* (Doney et al., 2004) highlighted four fundamental science questions to address for the oceans:

- What are the global inventory, geographic distribution, and temporal evolution of anthropogenic CO_2 in the oceans?
- What are the magnitude, spatial pattern, and variability of air-sea CO_2 flux?
- What are the major physical, chemical, and biological feedback mechanisms and climate sensitivities for ocean organic and inorganic carbon storage?
- What is the scientific basis for ocean carbon mitigation strategies?

In 2005, the U.S. Carbon Cycle Science Program established the Ocean Carbon and Climate Change (OCCC) program and the OCCC Scientific Steering Group (OCCC-SSG) to address multi-agency coordination of carbon cycle research. In February, 2006, the National Science Foundation (NSF), National Aeronautics and Space Administration (NASA), and National Oceanic and Atmospheric Administration (NOAA) decided to combine several related ocean programs, including OCCC, the Surface Ocean Lower Atmosphere Study (SOLAS), and the Integrated Marine Biogeochemistry and Ecosystem Research (IMBER) under one umbrella organization called the Ocean Carbon and Biogeochemistry (OCB) program to more broadly address issues of marine biogeochemistry (including carbon) and associated ecology.

Through the coordinated efforts of U.S. funding agencies, the NACP and OCB programs have made substantial progress in addressing the first two goals of the 1999 Science Plan. Less progress was made on the latter goals, however, although some investments by funding agencies did occur. Relatively little progress was made on Goal 5 in particular (developing the scientific basis for societal decisions about management of CO_2 and the carbon cycle), but also to some extent Goal 3 (impacts of land use on the carbon budget) and Goal 4 (projections of future atmospheric CO_2 concentrations).

2.3 Other relevant developments since the 1999 Science Plan

This section summarizes some key additional reports that have been published over the last decade, and that have led to evolving priorities for carbon cycle research. As such, some readers may wish to jump directly to Chapter 3.

Since publication of the 1999 carbon cycle plan, several documents have revised components of the U.S. Climate Change Research Program and highlighted the need for a new U.S. Carbon Cycle Science Plan. Collectively these documents provide insight into the research needs and the effectiveness of the program as described in this Plan.

The committee on the Human Dimensions of Global Change of the National Research Council (Stern, 2002) hosted a workshop in 2001 that highlighted the need for greater research on the human dimensions of the carbon cycle. The group concluded that although the 1999 Science Plan "notes the critical role of human activities in perturbing the carbon cycle, it does not include any research on these activities. The U.S. government's carbon cycle research activity has not yet integrated the relevant fields of the social and behavioral sciences." The committee wrote of building bridges between the natural and social sciences to produce the understanding necessary to "inform public decisions."

The *Strategic Plan for the Climate Change Science Program* (Climate Change Science Program, 2003) was the first comprehensive update of a strategic plan for the U.S. Global Change Research Program (1989). Five research goals were identified to focus research and to synthesize knowledge around broad strategic questions. The six research questions for the Global Carbon Cycle (see Chapter 7 of the Strategic Plan) were derived from the five goals recommended by the research community in *A U.S. Carbon Cycle Science Plan* (Sarmiento and Wofsy, 1999).

In 2007, the National Research Council presented its first review of the progress since the U.S. Climate Change Science Program/U.S. Global Change Research Program was established in 2002 (NRC, 2007b). For Global Carbon Cycle questions (7.1–7.6) from the 2003 *Strategic Plan for the Climate Change Science Program,* the report declared

that good progress had been made in documenting and understanding the current carbon sources and sinks. The North American carbon budget had been recently assessed (see following paragraph; SOCCR, 2007), but improvements in the observation and modeling approaches were needed to reduce uncertainties (Q 7.1). Focused research efforts and the synthesis of decades of observations reduced the uncertainties associated with the size of the ocean carbon sink, but significant uncertainties in ocean carbon processes remained (Q 7.2). On the subject of land use change and the carbon cycle, the report concluded that good progress had been made in understanding the historical relationships between land use and the carbon balance, but great uncertainties in future land management scenarios limited predictive capability (Q 7.3). Although fair progress had been made in linking changes in regional and global rates of CO_2 accumulation to climatic anomalies, understanding of the processes underlying some of these relationships was poor and limited our ability to predict factors that will dominate in the future (Q 7.4). Prediction of future fossil fuel emissions as well as carbon sources and sinks associated with future changes in land management were limited by the lack of involvement of stakeholder communities (Q 7.5). Informing carbon management was a new area of emphasis for the carbon cycle research element and had not yet made adequate progress (Q 7.6).

Also in 2007, the Climate Change Science Program and Subcommittee on Global Change Research published *The First State of the Carbon Cycle Report (SOCCR): The North American Carbon Budget and Implications for the Global Carbon Cycle* (SOCCR, 2007). The report provided a synthesis and integration of the current knowledge of the North American carbon budget and its context within the global carbon cycle.

In 2008, a *Revised Research Plan for the U.S. Climate Change Science Program* provided an update to the 2003 *Strategic Plan for the Climate Change Science Program* to take into account the advances in the science and changes in societal needs, draw on the program's long range planning process, and comply with the terms of the 1990 Global Change Research Act (Climate Change Science Program, 2008). That document highlighted the progress and accomplishments of the carbon cycle research elements over the previous four years and encouraged the continued development of future plans.

In 2009, the National Research Council published the report *Restructuring Federal Climate Research to Meet the Challenges of Climate Change* (NRC, 2009b). This report describes a new framework for generating the scientific and socioeconomic knowledge needed to understand and respond to climate change. It identified six priorities for a restructured U.S. Global Change Research Program (USGCRP) that would help develop a more robust knowledge base and support informed decisions. The document recommended reorganizing the USGCRP around integrated scientific-societal issues; establishing a U.S. climate observing system; supporting a new generation of coupled Earth System Models; strengthening research on adaptation, mitigation, and vulnerability; initiating a national assessment of the risks and costs of climate change impacts and options to respond; and coordinating federal efforts to provide climate information, tools, and forecasts routinely to decision makers.

In 2009, the NRC also published the report *Informing Decisions in a Changing Climate* (NRC, 2009a). This report examines the growing need for climate-related decision support, that is, organized efforts to produce, disseminate, and facilitate the use of data and information in order to improve the quality and efficacy of climate-related decisions.

In August, 2009, a group of prominent U.S. scientists concerned about climate change and the global carbon cycle wrote an article in the American Geophysical Union newsletter EOS, entitled *Carbon cycle observations: Gaps threaten climate mitigation strategies* (Birdsey et al., 2009). The article appealed for robust and sustained carbon cycle observations. They noted that "key elements of a national observational network are lacking or at risk" and strongly urged a coordinated system of observations that includes satellites, such as Landsat, MODIS, and SeaWiFS (as examples of key carbon cycle tools vulnerable to loss), *in situ* observations, and direct atmospheric measurements to meet the needs of scientific understanding and mitigation policies.

Another USGCRP report published in 2009, *Global Climate Change Impacts in the United States* (U.S. Global Change Research Program, 2009b)*,* summarized the science and the impacts of climate change on the United States from 2009 and beyond. It was written to inform public and private decision making at all levels. It focuses on climate change impacts in different regions of the U.S. and on various aspects of society and the economy, such as energy, water, agriculture, and health. The report also highlights the choices we face in response to human-induced climate change.

In 2010, a suite of studies under an umbrella entitled *America's Climate Choices* was requested by Congress to examine the status of the nation's climate change research efforts and recommend steps to improve and expand current understanding (NRC, 2010a,b,c). The first report, *Advancing the Science of Climate Change* (NRC, 2010a), provides a compelling case that climate change is occurring and is caused largely by human activities. The second report, *Limiting the Magnitude of Future Climate Change* (NRC, 2010b), states that substantially reducing greenhouse gas emissions will require prompt and sustained efforts to promote major technological and behavioral changes. The third report, *Adapting to the Impacts of Climate Change* (NRC, 2010c), calls for a national adaptation strategy to support and coordinate decentralized efforts. Reducing vulnerabilities to impacts of climate change that the nation cannot, or does not, avoid is a highly desirable strategy to manage and minimize the risks.

The NRC also examined requirements for monitoring greenhouse gas emissions in a report entitled *Verifying Greenhouse Gas Emissions: Methods to Support International Climate Agreements* (NRC, 2010d). The report evaluated three categories of methods for estimating greenhouse gas emissions from individual countries: national inventories, atmospheric and oceanic measurements and models, and land use measurements and models. Recommendations for improving monitoring capabilities focused on near-term efforts that could be implemented by the United States over a period of three to five years.

Finally, in 2010, the NRC also published a report entitled *Ocean Acidification: A National Strategy to Meet the Challenges of a Changing Ocean* (NRC, 2010e). This document reviews the recent legislation regarding ocean acidification, examines the current state of knowledge, and identifies key gaps in information needed to help federal agencies develop a program to improve understanding and address the consequences of ocean acidification.

2.4 Successes and remaining challenges

Research pursued under the guidance of the 1999 Science Plan has: 1) established a consensus that there is a large Northern Hemisphere terrestrial sink but that we do not yet understand entirely where it is or the mechanisms controlling its variability, 2) determined that the oceans are a major carbon sink, but that the annual growth in that sink is unable to keep up with projected growth in annual CO_2 emissions, 3) acknowledged that we need to understand land use history in order to determine the present and future carbon budget, 4) affirmed that we need to improve projections of the future behavior of the global carbon cycle, and 5) contributed importantly to a developing archive of critical, long-term observational data.

Many of the research goals in the 1999 Science Plan remain important challenges for the coming decade. New research thrusts are also needed, however, and we characterize these thrusts in more detail throughout this document. The new thrusts include the need to evaluate uncertainties in the mechanisms controlling the carbon cycle, to understand the role of humans as agents and potential managers of change, to understand direct impacts of CO_2 on ecosystems, to improve coordination and collaboration of researchers from different scientific disciplines and from different geographic areas, and to more effectively address societal concerns. By sustaining research efforts from the last decade and including new priorities for the next, we will make progress in the basic sciences and provide stronger scientific input to decision makers for carbon cycle management decisions.

Chapter 3
Fundamental Science Questions

Expanded research on the global carbon cycle is needed to improve basic understanding of the Earth's physical environment, ecosystems, and biodiversity and of its coupled natural and human systems. Sustained observations, experiments, and analyses are required. The research agenda should identify areas with inadequate knowledge and outline the science needed to guide important decisions now and to prepare for the needs of future decision makers.

Given the background and history provided in Chapters 1 and 2, we define three overarching questions that guide this new U.S. Carbon Cycle Science Plan:

Question 1. How do natural processes and human actions affect the carbon cycle on land, in the atmosphere, and in the oceans?

Question 2. How do policy and management decisions affect the levels of the primary carbon-containing gases, carbon dioxide and methane, in the atmosphere?

Question 3. How are ecosystems, species, and natural resources impacted by increasing greenhouse gas concentrations, the associated changes in climate, and by carbon management decisions?

Although these three questions have some overlap, they are intended to provide primary focus respectively on 1) ongoing changes in the global cycling of carbon; 2) human decisions that influence the carbon cycle, whether as a conscious goal or inadvertently as a result of other objectives; and 3) the effect of increasing atmospheric CO_2 and CH_4 on ecosystems, biodiversity, and natural resources. Many of the challenges in

carbon cycle science for the upcoming decade will require that these three questions be examined in a coordinated fashion. We provide two examples here.

First, the carbon cycle science community needs to contribute more directly to efforts to predict future changes in climate. Uncertainty in the fate of the carbon 'sinks' in the natural components of the carbon cycle represents one of the three fundamental uncertainties in understanding future climate, together with the uncertainties in the feedbacks in the physical climate system and uncertainties in the future trajectory of anthropogenic carbon emissions. Examining the three questions outlined above will help to inform efforts such as the assessment reports led by the Intergovernmental Panel on Climate Change (IPCC). Predictions of atmospheric CO_2 and CH_4 are used to drive predictions of climate change. The IPCC and U.S. National Assessments of climate change rely on multiple climate models run with a common set of forcings, including scenarios for atmospheric CO_2 and CH_4. Developing realistic CO_2 and CH_4 scenarios depends on understanding processes controlling their variability, understanding the impact of policy and management decisions, and assessing impacts on ecosystems, species, and natural resources, as outlined in the questions above. In addition, error bounds and confidence limits for these estimates are critically important for establishing the uncertainty in climate change projections.

Second, many of the greatest uncertainties in the carbon cycle lie at the interface of the human and natural systems. Urban-scale studies can be used to examine the impact of policy and management decisions, within the context of a changing natural carbon cycle and human-managed ecosystems.

3.1 Question 1: How do natural processes and human actions affect the carbon cycle on land, in the atmosphere, and in the oceans?

This question supports an expansion of the process-oriented and diagnostic studies of air, land, and water that have been part of U.S. carbon cycle science for more than a decade. It also focuses increased attention on the effect of human actions on the carbon cycle, including new and collaborative studies needed to address the socioeconomic processes controlling anthropogenic carbon emissions.

A process-level understanding of the carbon cycle is needed to assess and anticipate changes in CO_2 and CH_4 fluxes and atmospheric concentrations. Process-level understanding is needed to develop credible projections of the future carbon cycle and climate, as increasing greenhouse gas concentrations are currently the primary driver of climate change. Significant progress has been made in developing global and regional carbon budgets over the last decade, but many fundamental science questions remain. Particularly important for the coming decade are an improved understanding of thresholds and tipping points in the Earth's carbon cycle; expanded studies of dynamic regions such as the tropics and boreal zones on land and the high latitudes and coastal regions in the oceans; studies of historical changes through examination of paleo and geologic records; and an improved understanding of the processes involved in human impacts, such as disturbance of terrestrial ecosystems, on the Earth's carbon cycle. There is a need to reconcile the land and ocean sources and sinks with atmospheric observations.

The research needed to answer this first question encompasses the role of humans in the Earth's carbon cycle, including the burning of fossil fuels and land use change and land disturbance. The economic, political, cultural, and behavioral processes governing anthropogenic carbon emissions are an essential element of this U.S. Carbon Cycle Science Plan.

This challenging research area is rich in fundamental science questions and critical to the management and policy decisions of the coming decades.

With its increased focus on the effects of human actions on the global carbon cycle, the first question reinforces the need for enhanced natural science research on the carbon cycle, as well as social science research, but it also broadens carbon cycle research to develop truly integrated research programs that include collaborations between natural and social scientists.

3.2 Question 2: How do policy and management decisions affect the levels of the primary carbon-containing gases, carbon dioxide and methane, in the atmosphere?

This second overarching question seeks to understand how policy decisions and management choices alter atmospheric CO_2 and CH_4 concentrations and the Earth's climate and ecosystems. These policy and management choices include intentional manipulations of CO_2 and CH_4 and cases where these impacts are incidental to other objectives. Important management choices include adjustments in the balance of renewable and fossil fuels for energy production, investment in carbon sequestration projects, the development of economic incentives for increased production of biofuels, the development of public transport, and the protection of agricultural soils. The rate and magnitude of changes in the carbon cycle today reflect management choices made over the last two centuries. Reducing CO_2 and CH_4 emissions will require policy and management choices that are scientifically, economically, and ethically sound and politically and technically feasible.

Policy choices and carbon management incentives should ideally reduce atmospheric concentrations of CO_2 and CH_4 while minimizing the risks of undesired side effects on the Earth's ecosystems, resources, and people. Effective decisions, however, are hampered by limited understanding of the impact of these decisions on people and the environment, the effectiveness at reducing greenhouse gas concentrations, and the likelihood and magnitude of adverse side effects. The carbon cycle science community needs to provide a quantitative and credible understanding of the impact of different policy decisions and management choices.

3.3 Question 3: How are ecosystems, species, and natural resources impacted by increasing greenhouse gas concentrations, the associated changes in climate, and by carbon management decisions?

Increasing atmospheric concentrations of CO_2 and CH_4 are fundamentally altering marine and terrestrial systems and could compromise the rich diversity and multitude of services these ecosystems provide. The impacts of increased greenhouse gas concentrations on ecosystems go well beyond changes in carbon storage, which is the focus of Question 1. For instance, some ocean biota and marine resources are likely to be harmed in the coming decades as a result of rising CO_2 and ocean acidification (Doney et al., 2009). Increased atmospheric CO_2 concentrations can alter ecosystem structure in terrestrial systems and the competitive balance among plant species through stimulation of photosynthesis (Gill et al., 2002; Körner, 2009). On land and in the oceans, the direct impact of increasing atmospheric concentrations of CO_2 and CH_4 on species and ecosystems is an important research need.

It is difficult to decouple research on the direct effects of atmospheric CO_2 and CH_4 on ecosystems from the impacts of climate change caused by greenhouse gases on the same systems. Research on the impacts of climate on ecosystems goes well beyond the bounds of this Plan. Nonetheless, given that ecosystems respond to the combined stresses of changes in climate and changes in atmospheric composition and also that ecosystems alter atmospheric composition and climate, it is particularly important to conduct research that considers the interplay of both factors.

To fully address Question 3, the carbon cycle science community must be linked actively with the climate and ecosystems research communities. These connections will lead to a more holistic research program that considers a suite of environmental parameters, including biodiversity and ecosystem health, water resources, land disturbance, non-carbon greenhouse gases, resource economics, human health, and physical climate issues. Where basic research on such environmental parameters reaches beyond the scope of a carbon cycle research plan, collaborative and cooperative research is essential. Carbon cycle research is vital to understand how ecosystems, species, and natural resources will be impacted by increasing greenhouse gas concentrations, climate change, and carbon management.

3.4   The critical role of observations

The field of carbon cycle science depends on a well-designed, well-executed, and carefully maintained observational system. In support of all the research goals outlined in this plan, an optimally designed and integrated system is needed for long-term observations, data collection, and data management.

Such a system should capture the atmospheric, oceanic, biologic, demographic, socioeconomic, geologic, and paleo-data needed to establish baselines, measure changes, understand processes, and evaluate mitigation activity (e.g., Houghton, 2007). Important socioeconomic data need to be collected along with traditional carbon cycle measurements. We need to not only establish the patterns of land use and land cover, but of land management, land control, and patterns of fragmentation. We need to understand not just the chemistry of the atmosphere, surface waters, and soils, but also the details of energy production, consumption, trade, and cost. Attention is needed to integrate systems of complementary data for the study of complex, interdisciplinary systems. These data are critical for tracking the global carbon system and for providing a record of the variability in the major pools of carbon and their controlling processes. Only through the availability of sustained observations can we construct and evaluate models that diagnose carbon fluxes, attribute their variability to underlying processes, and predict their behavior as the geochemical and climate systems change. We also need the capacity to respond promptly to measure and evaluate short-term events such as extreme storms, large fires, radioactive releases, or economic dislocations, in order to take advantage of the insights that can be gained from such events. We need to commit to sustained data systems, while acknowledging the value of creative, single-investigator observations.

The 1999 U.S. Carbon Cycle Science Plan emphasized objectives for establishing observational networks and experimental manipulations that were only partially achieved. In general, the U.S. research community has excelled at developing and testing innovative observational and experimental methods and facilities. New observing systems and networks have emerged, but the networks have not always been maintained and expanded in a coordinated fashion, their long-term continuity has not always been ensured, the density of observations is irregular globally, and the types of observations collected have not been selected in a coordinated fashion to cover the necessary information. For example, many manipulative experiments have been conducted, but great uncertainty remains about the number, type, location, and longevity of experiments that are needed to satisfy the science questions articulated in the 1999 Science Plan. The focus on innovation must be maintained and our ability to expand long-term observations and to maintain proven, essential data sources must be strengthened.

A detailed listing of observational needs is beyond the scope of this Plan. However, Birdsey et al. (2009) have discussed many pressing observational issues facing the carbon cycle science community today, and general recommendations will be discussed in Section 5.1 of this document. For example, satellite observations of the land and oceans are a fundamental tool for carbon cycle research, but the continuity and quality of these observations is threatened. Multiple means of observations and experiments from multiple U.S. agencies are essential, and both cross-agency and international cooperation and coordination will be critical to the success of carbon cycle science in the coming decade.

Data management and open data access must also be high priorities of a successful observing network in the coming decade. Unlike weather data, which is primarily used shortly after being collected, data on carbon and climate tend to become more valuable with time. The archiving, management, documentation, and access to data in consistent, compatible, and easily accessible formats need to be carefully planned and thoughtfully implemented.

3.5   Dealing with uncertainty

In balancing the global carbon cycle and gaining a process-level understanding of its components, it is important to evaluate, understand, and deal with the uncertainty that arises through measurements, models, analyses, and projections. The complexity of the global carbon cycle dictates that few scientists will be expert in the full range of enquiries (e.g., Donner et al., 2009) and this emphasizes the importance of clearly conveying what is known and the uncertainty attached to that information. Uncertainties in the carbon cycle cause large uncertainties in future changes in the climate system, and, conversely, uncertainties about future climate make future changes in the carbon cycle more difficult to predict. As carbon cycle science moves into the public consciousness and stimulates political and economic decisions, knowledge of uncertainty is increasingly important. Public opinion polls reveal confusion about the existence of anthropogenic climate change and a lack of common understanding. For both scientists and the broader public, there is a need to address the sources and kinds of uncertainty.

Although error bounds are sometimes, though rarely, reported in current studies, these 'uncertainties' typically only represent the sensitivity of the estimates to a few factors. Instead, working with incomplete knowledge requires tools to quantify the full uncertainty of estimates to ensure that the truth lies within stated uncertainty bounds, and approaches to deal with the uncertainty that remains. For example, fossil fuel emissions are among the best-constrained components of global emissions, and yet the uncertainties are large enough to impact our efforts to balance the global carbon budget. Trading of emissions permits and verification of emission limitation agreements are two areas in which proper quantification and consideration of uncertainty have economic and political implications.

Characterization of uncertainty can be improved through additional observations and through modeling. Model estimates are impacted both by parameter uncertainty and by uncertainty in model structure, and inter-model comparisons can be useful in helping to understand these sources of uncertainty. As approached in recent IPCC publications, uncertainty can be conveyed as quantified measures expressed in probabilistic terms or as qualitative statements of the level of confidence based on the type, amount, quality, and consistency of evidence. Many documents have been published on the subject of how uncertainty should be quantified and clarified (e.g. Enting, 2008; Donner et al., 2009; Morgan et al., 2009). The U.S. community should work toward a more standardized approach for assessing uncertainty and improving its ability to clearly communicate these assessments to a broad audience.

Assessments are needed to determine what levels of 'certainty' are required for science and for management decisions, and what the possibilities and costs are for reducing uncertainty. For example, uncertainties currently limit accurate mapping of carbon fluxes at regional scales, which is the scale needed to verify emissions of individual countries. In addition, emissions offsets are likely to involve components of the carbon cycle that currently have very different levels of uncertainty. In many cases there is still a need to reduce uncertainty through targeted research, but in other cases it may be necessary or sufficient to simply be clear about the uncertainty in understanding that we do have. We need to ask about the cost of reducing uncertainty and social utility of doing so.

Chapter 4
Science Plan Goals

The research goals outlined here leverage the baseline provided by the 1999 plan, and many of the components of that initial plan remain important. Given research progress since the 1999 plan and newer research challenges that have arisen, however, we propose several new directions for the coming decade. One initiative is to devote more attention to human aspects of the carbon cycle, including the influences of social, political, and economic processes. Although human dimensions were mentioned in the 1999 plan, the social and natural science components of carbon cycle research have not been well integrated. Another new component is the study of the direct effects of increased CO_2 on ecosystems. The 1999 plan focused primarily on carbon accounting. It did not adequately address issues such as ocean acidification and restructuring of terrestrial ecosystems that can have a dramatic impact on biodiversity and human food supply, independent of climate change. A third new direction is to expand the carbon science program to include research that is more responsive to the needs of decision support. In particular, scientists need to understand the effectiveness of potential carbon management strategies (and the carbon cycle impact of management strategies pursued in support of other objectives) to inform decision makers of the full consequences of such management. Recognizing that public policy and human actions are being motivated by concerns about global climate change, we also need to make a greater effort to understand and convey the uncertainty associated with our knowledge of the carbon cycle.

These new directions are reflected in the fundamental science questions described in Chapter 3, which provide principles for guiding carbon cycle research. These questions, however, are unlikely to be answered completely in the next decade. To make progress in answering the questions and to provide guidance for human actions, we have outlined six science goals that should be addressed over the next decade, with proper funding and collaboration. These goals are listed here, together with references to the overriding questions they are primarily designed to address:

Goal 1. Provide clear and timely explanation of past and current variations observed in atmospheric CO_2 and CH_4 – and the uncertainties surrounding them. (Q1, Q2)

Goal 2. Understand and quantify the socioeconomic drivers of carbon emissions, and develop transparent methods to monitor and verify those emissions. (Q1, Q2)

Goal 3. Determine and evaluate the vulnerability of carbon stocks and flows to future climate change and human activities, emphasizing potential positive feedbacks to sources or sinks that make climate stabilization more critical or more difficult. (Q1, Q2, Q3)

Goal 4. Predict how ecosystems, biodiversity, and natural resources will change under different CO_2 and climate change scenarios. (Q3)

Goal 5. Determine the likelihood of success and the potential for side effects of carbon management pathways that might be undertaken to achieve a low-carbon future. (Q1, Q2, Q3)

Goal 6. Address decision maker needs for current and future carbon cycle information and provide data and projections that are relevant, credible, and legitimate for their decisions. (Q1, Q2, Q3)

One concern when dealing with a multidisciplinary problem such as changes in the global carbon cycle is the need to draw boundaries for research, especially because no research plan can address all facets of the problem and because research projects and communities with intersecting needs already exist. The challenge is to embrace new directions within the carbon cycle plan and to foster strengthened collaboration with other vibrant research communities with common or supporting interests. While the boundaries that we describe are not perfectly sharp, the motivation and scientific directions envisioned for each goal are further developed in the following sections.

4.1   Goal 1: Provide clear and timely explanation of past and current variations observed in atmospheric CO_2 and CH_4 – and the uncertainties surrounding them.

Do we understand the processes behind observed changes in the atmospheric concentrations of CO_2 and CH_4? Are these concentrations changing in predictable ways in response to our mitigation initiatives? The scientific community needs to provide to the public a clear and timely explanation of past and current variations observed in atmospheric CO_2 and CH_4, as well as the uncertainties surrounding these explanations. We note that 'timely' is an important part of this goal: to serve public policy needs, atmospheric observations and clear analysis are needed in close to real time. To address this goal, we need to develop the capability to estimate variability in carbon sources and sinks as well as the processes controlling that variability.

4.1.1   Motivation

Understanding historical and contemporary variations in the carbon cycle is vital for science and society. Modern variability of trace gas distributions from the global observational network provides a direct record of anthropogenic influence on the chemical composition of the atmosphere, ocean, and biosphere, while also demonstrating the potential of natural phenomena such as El Niño and droughts to impact the current and future climate. Contemporary observational efforts, process studies (including experimental manipulations) and modeling are required for verifying surface flux estimates, including those of fossil fuel emissions, terrestrial biosphere uptake, biomass burning and land use change and management, and the global ocean sink. Process-based understanding also makes it possible to predict future

variations in the efficacy of natural terrestrial and oceanic carbon sinks. Where atmospheric increases are tied to human activities, it is important to understand the link between human activities and atmospheric concentrations and that any attempt to purposefully manage carbon emissions is having the expected outcome. Finally, sustained observations and modeling efforts provide an early warning capability for detecting possible changes, such as methane release from Arctic tundra or other unanticipated effects.

4.1.2   Progress over the last decade

Contemporary records of atmospheric and marine chemical composition have proven vital for verifying human impacts on the Earth system, as have records of fossil fuel consumption and land use change. Of these records, the steadily increasing concentration of CO_2 in the atmosphere revealed by the Mauna Loa Keeling curve is perhaps the most well-known and clearest evidence of anthropogenic change. Records such as these have been used to infer the importance of the land biosphere and the ocean in modulating atmospheric CO_2, including showing that the atmosphere has historically retained only about half of anthropogenic CO_2 inputs, that the ocean has switched from a net source of CO_2 in preindustrial times to a net sink absorbing 25 to 30% of the annual anthropogenic CO_2 emissions, and that the terrestrial biosphere of the northern extra-tropics currently represents a significant carbon sink. These long-term records have revealed the importance of the terrestrial biosphere in influencing atmospheric composition, and they have similarly shown that large-scale climate modes (e.g., El Niño), volcanically-injected aerosols, and regional-scale precipitation and temperature anomalies can all leave detectable fingerprints on terrestrial carbon cycling. Time series of meteorological and biological observations from tower flux networks have revealed the importance of disturbances from harvest and fire on terrestrial carbon processes and thus sources and sinks (Amiro et al., 2010), as well as interannual variability in carbon and water processes associated with drought and early spring warming.

In the ocean, large-scale carbon surveys, time-series stations, and observations from instruments placed on commercial ships have shown that CO_2 concentrations in the surface ocean are increasing at about the same rate as in the atmosphere, but large-scale circulation patterns are limiting the rate at which that CO_2 is moved into the ocean interior. Observed heterogeneity in the regional pattern of rising ocean pCO_2, however, indicates changing physical and biogeochemical processes, suggesting that these changes may cause future ocean uptake of CO_2 to depart from its historical trend. Understanding the nature of these changes is critical for future development of ocean carbon cycle models and thus for making accurate predictions of feedbacks that may alter the ocean's uptake of CO_2. Ocean observations have also demonstrated that ocean carbon uptake is strongly influenced by climate variability over a range of time scales as well as by long-term trends in changing ocean chemistry.

4.1.3   Major uncertainties

Significant uncertainties remain about what processes cause the observed changes in the atmospheric and oceanic composition of CO_2 and CH_4. For instance, we cannot reliably quantify the relative importance of CO_2 fertilization and land use changes for the net terrestrial carbon sink in the Northern Hemisphere or in the tropics. Observations from repeat hydrographic surveys and process studies have revealed variability in the ocean interior that challenges our understanding of ocean circulation and biology. The connections between the land, ocean, and atmosphere in coastal zones are still poorly understood. Providing timely explanations of observed variations, with robust attribution to natural and anthropogenic causes, as well as communicating that understanding to decision makers and the general public, are thus important goals of the current plan.

4.1.4   Scientific directions

Establish a continuity plan and continue expansion of carbon observing networks

The ability to explain variations in atmospheric greenhouse gas concentrations is predicated on the ability to resolve the spatial and temporal gradients of the gases. Three- to five-year funding cycles typical of scientific investments are inadequate for capturing the long time scales and large spatial scales needed for observing trends in greenhouse gas dynamics. To achieve Goal 1, the research community must continue to promote international cooperation and establish an alternative model for supporting long-term observations.

In addition, current carbon observation networks have been unable to resolve competing processes and their net impact on atmospheric CO_2 and CH_4 concentrations. Filling this gap will require a systematic and coordinated expansion of global carbon observations. This expansion will inevitably involve *in*

situ atmospheric concentration measurements, as well as an expanded array of greenhouse-gas observing capabilities from space. It should also involve additional *in situ* observations on land and in the ocean to better characterize the processes controlling carbon exchanges with the atmosphere and to help reconcile the differences between bottom-up and top-down flux estimates. Several of these efforts are already underway, including the continuing expansion of the NOAA Earth System Research Laboratory (ESRL) Cooperative Air Sampling Network, the ongoing development of space-based missions such as Orbiting Carbon Observatory 2 (OCO-2) and Active Sensing of CO_2 Emissions over Nights, Days, and Seasons (ASCENDS), the growth of the AmeriFlux network to more than 80 sites, and NOAA's ocean carbon-observing network. To ensure that this expansion is both sustainable and optimal, however, the carbon cycle science community must establish unprecedented coordination, both internationally and domestically among U.S.-based federal agencies, in identifying the best set of activities aimed at ensuring sustained observations.

Conduct manipulative experiments and process studies to provide mechanistic understanding of responses and feedbacks to changing greenhouse gas concentrations and climate

Biologic systems are subjected to a variety of interacting stresses, and controlled experiments and process studies often provide the best way to identify how these stresses operate alone and together. For instance, experiments on the role of changes in atmospheric chemistry, temperature, precipitation, and pathogens in terrestrial ecosystems provide the basis for modeling anticipated environmental changes. Sustained observations that include component processes and total ecosystem fluxes along disturbance and climatic gradients can also provide mechanistic understanding of responses to climate and disturbance.

Ongoing changes in the physical and chemical environment of the ocean are altering both the biological pump and the solubility pump in ways that can be described qualitatively but which cannot be coded in models because of insufficient understanding of the sensitivity of key processes to ongoing changes. These sensitivity factors must be constrained quantitatively before models can be used to make meaningful predictions of future ocean uptake of CO_2.

Ongoing changes in ocean stratification, winds, sources of micronutrients, acidification, supply of mineral ballast, land-ocean exchange, carbon and nutrient ratios of sinking organic matter, etc., are all affecting ocean biogeochemical cycles and marine ecosystems. Perturbation of ecosystems, in turn, impacts the efficiency of the ocean's biological pump, both through altered nutrient utilization efficiency and in changes in the transmission of sinking organic material through the mesopelagic zone. These are critical uncertainties that must be examined through a combination of sustained observations and process studies.

Develop models capable of constraining process-based understanding of carbon flux variability

Whereas much effort over the past decade has focused on quantifying sources and sinks of CO_2 and CH_4 at increasingly fine spatial and temporal scales (i.e., diagnosing the variability in the carbon cycle), efforts at attributing this variability to underlying biogeochemical and socioeconomic processes is still in its early stages. Therefore, an achievable result for the upcoming decade is to develop models capable of constraining the process-based understanding of observed variability. Such improved models will complement efforts to quantify fluxes. Whereas the net global carbon flux can be quantified very precisely using even the current observational network, and while fluxes at sub-kilometer scales can be measured more directly, the discrepancy in reported fluxes at regional and smaller intermediate scales is still very large, hampering our ability to assess net carbon budgets from cities to countries. These discrepancies include high uncertainty for issues as simple as separating the net oceanic and terrestrial fluxes, as well as disentangling biospheric from anthropogenic terrestrial fluxes.

4.1.5 Related Issues

The global carbon cycle does not operate in isolation. It is intimately linked to 'other' global cycles (e.g., water, nitrogen, phosphorus, oxygen) that must be studied to fully understand the carbon system. It is beyond the scope of this Plan to identify the aspects of these other cycles that are most critical for meeting Goal 1, but it is anticipated that a wide field of related studies will need to be linked to the carbon cycle research to round out our understanding of the controls on the global carbon cycle and to accomplish the stated goals. This goal requires involvement of research on the human

dimensions of the carbon cycle, because understanding the economic and policy effects on anthropogenic emissions and sequestration efforts is essential for understanding variations in atmospheric CO_2 and CH_4. In efforts to mitigate increases in atmospheric greenhouse gases, a process-based understanding is essential for how human activities impact greenhouse gas emissions and for monitoring and verifying that atmospheric concentrations are in fact exhibiting the expected outcomes. Motivating human mitigation activities could be very difficult without a compelling attribution of consequences. In addition, achieving this goal will require close collaboration with many global observational efforts related to the carbon cycle. Measurements of trace gas concentrations will remain crucial to verifying bottom-up estimates of fossil fuel emissions and land use change, and may also be called upon to evaluate the efficiency of sequestration and other carbon management strategies.

To make the research developed in Question 1 accessible to public and private decision makers, carbon cycle scientists will need to develop more meaningful carbon system metrics that can be explained to the broader public and tracked through time. New advances in understanding will need to be captured from the scientific literature and made accessible to the broader climate change community. In this vein, we recommend and anticipate an increasing emphasis on incorporating scientific understanding of the carbon cycle, as well as observations and model predictions, into integrated assessment models and other research tools used to evaluate and guide policy choices.

4.2 Goal 2: Understand and quantify the socioeconomic drivers of carbon emissions, and develop transparent methods to monitor and verify those emissions.

This goal seeks to derive process-level understanding of the human factors that determine carbon emissions from energy use, industrial activity, and land use and will require close collaboration with other research efforts such as those in integrated assessment. In particular, there is a strong need to characterize and understand the relative importance of key drivers of emissions regionally and over different temporal and spatial scales. It is important to improve our understanding of the multitude of human objectives and the relationships among carbon emissions and other human priorities and motivations, and how policies with very different objectives may affect carbon emissions. Improved understanding will enable policymakers to better deal with current emissions levels and provide better projections of future emissions from specific sources, including the implications of alternative policy scenarios. This research is timely not only as an essential component of understanding the evolution of the carbon cycle but also as an input to current policy debates regarding the role of reducing emissions from deforestation and degradation (REDD), the implications of expanded use of agricultural lands for biofuels, and tradeoffs between carbon and other societal goals such as food production and sustainable livelihoods.

As international and intra-national agreements to limit greenhouse gas emissions emerge, it is important to be able to independently measure, monitor, and verify reported emissions. Inventories of greenhouse gas emissions have traditionally been self-reported by countries, communities, or companies based on survey data on the activities that generate emissions and coefficients that convert these to emissions estimates. With international commitments or economic incentives involved, methods and systems are needed for evaluating emissions and the impact of carbon management strategies. Atmosphere-based measurements, remotely sensed observations, evaluation of socioeconomic parameters, and other tools need to be developed to provide confirmation and confidence in mitigation commitments. The institutions and infrastructure for monitoring and verification must come from the national and international political processes, but the tools and methods need to be developed by science.

4.2.1 Motivation

Human activity is now the dominant factor driving changes in the carbon cycle and its impact is expected to grow throughout the 21st century. Understanding future changes in the carbon cycle must therefore include study of the key drivers of emissions from human activity, whether from energy use, industry, or land use. Providing relevant information on the carbon cycle must include information on key human processes and drivers, as the effectiveness of many policies will depend on how human activities interact with these drivers. The effectiveness of policies will also depend on confidence that others are doing their agreed share. Although much can be done with self-reported emissions inventories and easily implemented improvements in those inventories, independent

methods are needed to improve and support self-reported estimates of emissions. One important role of the carbon cycle science community is to develop the tools, observations, and models that can be used to evaluate emissions.

4.2.2 Progress over the last decade

Research over the past decade has led to significant advances in understanding factors affecting anthropogenic carbon emissions, including the role of land use change, changes in consumption patterns, agricultural practice, urban development, and international trade. A growing number of economic and technological analyses of carbon management and conservation strategies have been aimed at limiting anthropogenic emissions. These efforts have been conducted by various researchers and research groups, including those who study land use and land cover change (e.g., flux networks, regional integration of *in situ* observations with terrestrial process models, land us and land cover change (LUCC), and the Global Land Project), urban form and metabolism (i.e., Urban and Global Environmental Change (UGEC)), technological development, energy resources, and integrated assessment modeling. There is an emerging body of work on cities, behavior change, and the willingness to pay for carbon offsets or to change various types of behavior in response to climate change. While much of this work has taken place outside of the carbon cycle science community as traditionally defined, tremendous opportunity exists to link, integrate, and enlarge the scope of research on drivers affecting carbon emissions and uptake.

Through efforts of the IPCC and national, industry, and nongovernmental organization (NGO) groups, marked improvements have been seen over the last 15 years in developing and implementing standardized methods for greenhouse gas emissions inventories. Current best practice in developed countries is able to produce reasonably accurate estimates of emissions of CO_2 (NRC, 2010d). Capacity building is required to extend these inventories to many developing countries. In addition, only limited data exist for independently evaluating many components of these inventories. New and transparent approaches for cross-checking inventories are needed to build confidence in mitigation agreements (NRC, 2010d). Land- and space-based approaches to support inventories or to falsify some inventory components must be developed, along with evaluation and modeling of related demographic and economic data.

4.2.3 Major uncertainties

Though we have greatly improved our basic understanding of the socioeconomic drivers of anthropogenic emissions and carbon use, many uncertainties remain. For example, there is still significant uncertainty in current emissions, particularly emissions from land use and for anthropogenic emissions at sub-national and sub-annual scales. In terms of a process-level understanding of drivers, advances have been made within particular foci, such as energy analysis, land use change, urban footprint analysis, and international trade. A high priority now is to understand interactions among these drivers and their relationship to meeting basic human needs and aspirations. An integrated systems understanding of the interactions among socioeconomic, policy, and cultural factors at different spatial and temporal scales will allow development of more effective mitigation and adaptation strategies. How socioeconomic drivers of emissions and uptake interact with the biophysical components of the carbon cycle is only beginning to be explored. As a prominent example, a substantial amount of research effort in the carbon cycle and climate communities will be based on the Representative Concentration Pathways (RCPs) of future emissions, developed for use in the IPCC assessment process. Development of socioeconomic scenarios, including emissions drivers and their interactions, is at an early stage and will require substantial new research.

Explorations with inverse modeling and preliminary studies of airborne and satellite measurements are beginning to establish relationships between CO_2 fluxes and observations of its atmospheric concentration. We do not yet know with certainty what kind of surface measurement system and modeling will be required for useful estimates of the emissions of cities, countries, or regions.

4.2.4 Scientific directions

In defining key scientific directions in the study of the motivations and drivers of anthropogenic carbon emissions and sinks, we draw on a sometimes useful typology that distinguishes direct from indirect drivers (Millennium Ecosystem Assessment, 2003). A driver is any natural or human-induced factor that directly or indirectly causes a change in anthropogenic emissions or alters carbon sinks. Direct drivers unequivocally influence emissions or sinks and therefore can usually be unambiguously identified and measured; indirect drivers operate more diffusely and typically affect emissions or sinks through their effects on direct drivers.

For example, direct drivers of emissions and influences on sinks include:

- The removal of fossil fuels from geological reservoirs on land and in the ocean and their subsequent refining and combustion
- Industrial processes, such as the production of cement, and some waste management processes that generate greenhouse gases
- Land use change that modifies terrestrial carbon stocks
- The purposeful sequestration of CO_2 in plants, soils, the ocean, or geologic reservoirs
- Processes leading to the emission of nitrogen gases and ozone precursors (which can affect the size and sign of terrestrial carbon sinks)

Indirect drivers include a wide range of human activities that influence direct drivers and therefore lead to changes in emissions or the operation of sinks, including the following key categories:

- Demography (population growth/decline, urbanization, aging, changing living arrangements)
- Economics (economic growth, wealth distribution, trade, job creation, and incentives for desired activities)
- Science and technology (research and development, technology diffusion, and their implications for energy supply, end use efficiency, agricultural productivity, sequestration methods)
- Legacy effects of land use, such as the time since deforestation or agricultural abandonment, which influence carbon uptake today
- Behavior (lifestyles, culture)
- Institutions (climate agreements, changes in markets, regulatory regimes, property rights)

Changes in public policy have a critical influence on drivers and therefore on emissions and sinks. For example, policies that place a value on carbon, promote technology development, or support agricultural production all provide incentives to change behavior and can have a strong influence on fossil fuel production and combustion and land use change.

Scientific progress on understanding how human activities influence the carbon cycle, and how these effects might evolve in the future, requires achieving the following objectives.

Quantify the relative importance of different socioeconomic processes and their interactions in different parts of the world and at different spatial and temporal scales.

Anticipating future carbon emissions and uptake requires understanding which processes are the key drivers in different places and over diverse time scales. Socioeconomic conditions and trends vary widely in different parts of the world. Economic growth is proceeding at different rates, with different levels of energy and carbon intensity and with varying incentives, opportunities, access to resources, physical setting, and vulnerability to climate change. Land use patterns also vary widely with factors such as demographics, climate, culture, history, and international trade playing various roles in different places (see, e.g., Lambin et al., 2001). As in many parts of this plan, interdisciplinary and international collaboration and cooperation will play a key role in achieving this objective.

Influences on the carbon cycle are driven not by any single socioeconomic factor, but by combinations of factors acting together. Understanding interactions among drivers is therefore crucial for anticipating future emissions. For example, we need research to understand relationships between economic development and demographics; linkages among economic growth, technological change, globalization, and energy systems; and how different types of policies will influence land use emissions. Finally, the relative importance of drivers will differ across spatial and temporal scales. The drivers of emissions at the level of an individual city, for example, can differ substantially from those driving emissions nationally, regionally, or globally. In addition, some processes, such as fluctuations in economic growth rates, may be important over a few years, other processes such as urbanization can be important over several decades, and still others, such as changes in energy supply technologies, are important over many decades to a century.

Better quantify the potential range of future emissions from energy and land use

Improving our ability to anticipate the future evolution of the carbon cycle will require improved projections of fossil fuel consumption/production and emissions and of human influences on carbon uptake. The ultimate goal of such projections is typically not to predict future conditions, but rather to better quantify the limitations and uncertainty in projections. In particular, an improved understanding of the plausible range of future emissions, land use, land disturbance, and influence on sinks will be critically important for informing policy and research related to the carbon cycle. Whether short-term fluctuations in emissions portend longer-term trends in emissions growth remains an open question (Raupach et al., 2007), and even estimates of current emissions from land use are highly uncertain. How rapidly the effect of human activity on the carbon cycle will grow and how rapidly it may be curtailed, are critical questions that require concerted interdisciplinary research. The necessary work will draw on improved data for current and past socioeconomic conditions and trends and improved models of the interactions of socioeconomic processes.

Determine how carbon prices and other policies affect socioeconomic drivers and emissions

With increasing attention to developing and implementing policies that mitigate human influence on the carbon cycle, understanding how policies, particularly those that price carbon, will affect the processes driving emissions is increasingly important. How will different policies influence the pace and direction of technological change, patterns of land use, and the consumption choices of individuals? How are changes in indirect drivers likely to influence policy decisions? How might alternative policy goals, including concentration or temperature stabilization goals, influence the types and timing of policies to be implemented? Progress on this goal will require concerted efforts to study the economics of carbon and interdisciplinary research linking economics, energy systems analysis, land use science, and models of the natural carbon cycle. Collaborative efforts with the integrated assessment and other research efforts are needed to ensure that comprehensive data and models serve this complex objective.

Develop the tools, observations, and models needed to quantify and evaluate emissions

Efforts to limit emissions will depend on being able to motivate and measure the success of those efforts and to instill confidence that objectives and commitments are being met. Gas concentrations, isotope signatures, trace species, land surface properties, and measures of socioeconomic drivers all provide data potentially useful for monitoring and verification of emissions. To reduce uncertainties associated with emissions from land cover change, high spatial resolution satellite observations, light detection and ranging (LIDAR), and inventory observations need to be sustained and integrated with other types of socioeconomic data. Airborne and satellite measurements of atmospheric CO_2 and CH_4 offer another dimension for constraining both energy and land use emissions. Instrument improvements are still required, however, and it remains unclear how much passive and active measurement systems will be able to contribute. Models need to be developed to accurately constrain emissions based on the information provided by these various types of measurements. Methods capable of constraining regional inventories at the scale of individual countries and states should be emphasized, because these scales are key to the implementation of carbon management policies. Standardized methods are needed for building and evaluating such inventories.

Systems that are appropriate for cooperative parties or countries and those that might be used for uncooperative parties need to be considered. International partnerships and collaborations will play a central role in establishing measurement systems and analyses that provide reliable and transparent results. The carbon cycle science community is in a unique position to develop the tools, observations, and models needed to quantify and evaluate emissions. However, it is important to distinguish the role of the carbon cycle science community in the scientific development of these tools, and the role of decision- and policymakers in monitoring and verifying emissions and compliance with carbon management strategies.

4.2.5   Related Issues

Considerable research is underway on aspects of the drivers of anthropogenic emissions and influences on uptake, including work by research communities in integrated assessment modeling (e.g., US DOE, 2007), urbanization (e.g., UGEC, 2010), energy systems (e.g., The Global Energy Assessment (IIASA, 2010)), the Global Land Project (GLP, 2010)), and the Global Carbon Project (GCP, 2010). Continuing commitment, improved coordination, strong collaboration, improved integration, and increased attention to the critical role of humans in the global carbon cycle are essential for improved understanding, more effective mitigation, and successful adaptation to a changing carbon cycle.

4.3   Goal 3: Determine and evaluate the vulnerability of carbon stocks and flows to future climate change and human activity, emphasizing potential positive feedbacks to sources or sinks that make climate stabilization more critical or more difficult.

All carbon reservoirs and carbon processes are not equally vulnerable to change, resilient to stress, responsive to management, or susceptible to unintended side effects of management decisions. We need to identify which carbon pools and flows are most vulnerable and to understand the physical, chemical, and biological processes important in determining the degree of vulnerability of these pools and flows. We also need to predict the consequences of carbon management and sequestration schemes on vulnerable pools and to support carbon management goals by prioritizing the most vulnerable stocks and flows and the resources that are needed to assure the stability of these.

Vulnerability also needs to be understood in terms of direct feedbacks to the climate system because, for example, changes in the physical and chemical characteristics of the oceans can impact the distribution of heat and moisture in the Earth system and changes in terrestrial carbon stocks can affect the land surface energy balance through changes in albedo and latent heat transfer.

4.3.1   Motivation

The carbon cycle will respond to climate change in major ways, but not all carbon reservoirs and carbon processes are equally vulnerable or resilient to stress. Vulnerable carbon pools may release a large amount of carbon, providing a positive feedback to climate change. Some changes may also be abrupt or not easily reversible (e.g., release of methane hydrates from continental shelf regions). Identifying vulnerable carbon stocks and flows, understanding the processes controlling their behavior, and evaluating the risk and magnitude of significant changes in their net impact on the carbon cycle is critically important in anticipating the degree of future climate change. We need to improve our skill and confidence in anticipating hot spots and tipping points of stocks and flows that are most vulnerable to future changes. Predicting the likelihood of substantial changes in vulnerable carbon stocks and flows is necessary for devising strategies for climate mitigation, adaptation, and carbon stock management. By far the largest 'active' pools of carbon on Earth are in the ocean and soils. Even if these pools are not currently considered the most vulnerable, it is important to continue to assess the processes controlling exchanges with these pools as even a small change in the net balance of exchanges could have a significant impact over time.

4.3.2   Progress over the last decade

Important progress has been made in identifying some vulnerable carbon pools in recent years, including carbon stored in high-latitude permafrost, peatlands in tropical and other regions, forests vulnerable to insect pests or wild fires, soils vulnerable to plowing and decomposition, freshwater and coastal wetlands, systems with methane clathrates on land and in the ocean, and ocean basins that are vulnerable to changes in biology or ocean circulation. Recent estimates have suggested that very large stores of carbon exist in some of these vulnerable pools, such as permafrost (Schuur et al., 2008). In the case of permafrost, large quantities of methane could be released if substantial amounts of permafrost melt as bacteria feed on the ancient carbon and nutrient stores at high latitudes (Walter et al., 2006). These stores are not well represented in existing models, and theoretical and modeling studies have suggested a range of possible ecosystem shifts in response to global warming (Schneider et al., 2007). For instance, carbon reservoirs in areas such as tropical rainforests are vulnerable due to the potential of future drought in the subtropics (Cox et al., 2000), and tropical peatlands are vulnerable to draining for land use change and subsequent carbon loss from fires after drought (Page et al., 2002). Coupled carbon-climate models suggest potentially large climate feedbacks due to changes in these vulnerable carbon reservoirs, leading to accelerated warming.

The ocean contains over 50 times more carbon than the atmosphere does and will ultimately, over millennia, absorb most of the fossil carbon released to the atmosphere. Estimates of the current uptake by oceans have greatly improved over the last decade. Although the oceans are not currently considered the most vulnerable carbon pool over decadal time scales, additional understanding of how future ocean uptake may change is still needed. Rapid changes are occurring in some marine systems, such as carbon storage and fluxes in coastal wetlands and waters (e.g., Cai, 2011), in coastal hypoxic zones, and in the high-latitude ocean margins and basins (e.g., Bates et al., 2009).

4.3.3   Major uncertainties

Uncertainties associated with changes in vulnerable carbon stocks and flows are still large (Schimel et al., 2001). Given that approximately half of the carbon released through fossil fuel burning is currently taken up by sinks on land and in the ocean, predicting the future carbon balance of these major reservoirs is critically important. One key uncertainty arises from how the carbon cycle interacts with climate change. Differences in the treatment of climate and CO_2 fertilization across general circulation models (GCMs) leads to projected differences of almost 300 ppm CO_2 in 2100 for identical fossil fuel emission scenarios, with a resulting uncertainty in surface temperature of approximately 2.5°C (Friedlingstein et al., 2006). Changes in the frequency or intensity of large-scale disturbances are another important uncertainty. Additionally, human activities and land management can change rapidly through economic incentives or global trade. Development of biomass energy is an example of how quickly grasslands, croplands, and other ecosystems can be converted based on policy or economic incentives. Large-scale thinning of forests for biomass energy/biofuels can result in significant changes in ecosystem processes and biodiversity as a result of policy or economic incentives.

An additional uncertainty is that all vulnerable carbon reservoirs and flows have likely not been identified or thoroughly assessed. The processes that control the ways in which carbon reservoirs on land respond to changes in temperature, soil moisture, and other factors or the ways in which the oceans respond to changes in carbonate chemistry or the health of marine phytoplankton and surface mixing are generally not well known. Even for known vulnerable carbon stocks, significant uncertainties remain. For example, permafrost carbon pools under tundra and on the continental shelves are ancient, yet gaps in our understanding remain for how they were formed and evolved under past changing climate and possibly human influence. One major challenge is the balance of competing effects and feedbacks that cannot be assessed without a holistic view. For instance, the potential loss of soil carbon at high latitudes may be countered by increased vegetation growth, and the net carbon loss or gain may be sensitive to the degree of warming (Qian et al., 2010).

4.3.4 Scientific directions

New research is needed to identify the most vulnerable carbon pools and flows and to study and model the processes that make them vulnerable and the potential consequences of this vulnerability.

Identify vulnerable pools and flows and monitor their changes, especially those that may change more rapidly in the near future.

Some carbon reservoirs and carbon processes are more vulnerable to changes in climate or carbon cycling than others. Moreover, carbon reservoirs and processes differ in their resilience to stress, responsiveness to management, and susceptibility to unintended side effects of management decisions. Research needs to focus on quantifying the known vulnerable pools and flows, tracking changes and rates of change in their size, and identifying and evaluating any additional vulnerable pools or flows. Studies of paleo records of past events will be critical for identifying and understanding these vulnerable pools. Long-term, sustained observational networks must include *in situ* atmospheric, flux tower, and remote-sensing observations and carbon stock assessments, as well as examination of historical records.

Understand the physical, chemical, and biological processes important in determining the degree of vulnerability of carbon pools and flows, and build such understanding into diagnostic and mechanistic models.

To anticipate future changes, and to plan for management actions, we need a thorough understanding of the processes underlying potential changes in vulnerable carbon pools and flows. Controlled experiments can play an important role in this effort because parameters can be manipulated to represent possible future carbon and climate scenarios so that vulnerability under extreme conditions and novel combinations of environmental factors can be tested. Strategic location of *in situ* terrestrial observations along ecotones and gradients, such as those associated with flux towers combined with biological measurements, and other areas already identified as potentially vulnerable, will play an important role in tracking changes in carbon pools and fluxes for calibrating mechanistic models. Diagnostic and mechanistic models then become critical in delineating and quantifying the relative roles of the processes controlling carbon balance in vulnerable reservoirs. The many processes involved in the carbon cycle demand collaboration and knowledge from an unprecedented number of traditional fields, and pose a major challenge in the management of scientific research. We must develop effective new ways to facilitate interdisciplinary and innovative research to address this need. The links to physical oceanography and surface land energy balances need particular attention.

Predict the likelihood, timing, and extent of potential changes in vulnerable carbon stocks and flows with numerical models and empirical methods.

Empirical methods are valuable for extending our knowledge of past changes into the future. Numerical models represent the biological, physical, and chemical processes controlling carbon balance and can be informed by available observations; they provide a key tool for predicting future changes in a mechanistic way. Therefore, we need to develop models to represent accurately the past, present, and future behavior of vulnerable carbon stocks and flows. Model inter-comparisons offer a useful opportunity to integrate knowledge across modeling platforms. This approach is especially important as vulnerability is often nonlinear, and abrupt changes may not be easily constrained by past short-term observations. Studies of past changes in carbon stocks can provide valuable insights from changes at geologic time scales or in earlier geologic times.

Predict the consequences of carbon management and sequestration schemes on vulnerable pools; support carbon management goals by helping to prioritize the most vulnerable stocks and flows that require management and the resources that are needed.

Comprehensive models will also be an essential tool in evaluating carbon management schemes, in order to avoid undesirable side effects of proposed management strategies. Such consequences may be difficult to anticipate due to the complexity of the carbon cycle, and carbon cycle models offer a tool for balanced evaluation. An exciting opportunity in using carbon cycle models is to identify the most vulnerable carbon stocks and flows. Once those vulnerabilities are clearer, we can test management possibilities and evaluate effectiveness in reducing vulnerability and altering the global carbon budget, including the resources needed for achieving these management goals.

4.3.5   Related Issues

Vulnerability in terrestrial carbon pools and fluxes is directly related to physical changes in ecosystems and climate, including changes in water resources, sea level, ocean circulation, energy, food supply, resource extraction, and livelihoods. Because abrupt changes in vulnerable carbon pools may be eye-catching, links to the public, media, and decision makers may be direct and prompt. On the other hand, vulnerability in marine systems may be more related to water, carbon, and energy flows and will be manifest in quite different ways. Public perceptions and management decisions need to be informed by process-based understanding and clear understanding and presentation of uncertainty.

4.4   Goal 4: Predict how ecosystems, biodiversity, and natural resources will change under different CO_2 and climate change scenarios.

Increasing concentrations of atmospheric CO_2 and other greenhouse gases have been and will continue to be a reality for the foreseeable future. The direct effects of elevated greenhouse gas levels, along with the accompanying changes in climate, are likely to alter ecosystems profoundly on land and in marine and freshwater environments. The interaction of climate change and the carbon cycle is of primary importance and this interaction is discussed in Goals 1, 2, and 3, recognizing that the ecosystem effects of climate change go far beyond the scope of this Plan. The specific focus of the goal presented here, therefore, is to focus on the *direct* impact of increasing atmospheric greenhouse gas concentrations on ecosystems, beyond their potential role as carbon reservoirs or sinks. Three examples of such impacts are altered marine ecosystem structure due to ocean acidification, biodiversity impacts on land, and the potential stimulation of net primary productivity due to additional CO_2.

4.4.1   Motivation

Atmospheric levels of carbon dioxide and other greenhouse gases are strongly mediated by terrestrial and aquatic ecosystem processes. Correspondingly, ecosystems are highly sensitive to changes in greenhouse gas levels, even in the absence of climate change. For instance, rising levels of atmospheric CO_2 and other greenhouse gases alter many ecological factors, such as the chemistry of surface waters and the biodiversity of terrestrial and marine ecosystems. These and other effects

have critical implications for society, including impacts on fisheries and agricultural production. Moreover, understanding these effects is critical to identifying potential feedbacks and thresholds in the interactions among ecosystems, climate, and atmospheric chemistry.

4.4.2   Progress over the last decade

Over the past decade, we have come to a better understanding of the profound ecosystem changes that have occurred with changing greenhouse gas concentrations and other climate-related forcings. Shifts in ecosystems due to changes in temperature, water availability, increased CO_2 levels, and other factors have altered biodiversity, ecosystem structure, and associated partitioning of carbon between land or ocean and the atmosphere (Denman et al., 2007; Field et al., 2007). In terrestrial systems, the range and phenology of many species are already changing in response to climate change (Root et al., 2004; Rosenzweig et al., 2008). In many parts of the world, future species composition is expected to differ substantially from that of today (Williams et al., 2007).

In coastal and marine ecosystems, rising sea level and intense coastal development have led to widespread loss of vegetated coastal habitats including mangroves, salt marshes, and seagrasses, negatively impacting carbon burial capacity and biodiversity (Duarte et al., 2005; Waycott et al., 2009). Alteration of seawater chemistry from excess CO_2 has been well documented and the resultant ocean acidification threatens coral reef ecosystems and other benthic and pelagic marine food webs, and could diminish both biodiversity and the effectiveness of ocean carbon sinks (e.g., Riebesell, 2008). Ocean acidification in coastal waters and estuaries threatens a seafood industry worth tens of billions of dollars. Some satellite observations suggest long-term declines in global ocean productivity related to climate (Behrenfeld et al., 2006) and an expansion of oligotrophic ocean waters presumably related to increasing ocean thermal stratification (Pörtner, 2008). A further consequence of the combined effects of rising CO_2 and ocean warming is an expansion of ocean dead zones (Brewer and Peltzer, 2009).

4.4.3   Major uncertainties

Major uncertainties remain in our understanding of how marine and terrestrial ecosystems respond to increasing greenhouse gas concentrations. For the oceans, recent findings indicate that ecosystem structure can substantially alter vertical export of carbon in ocean ecosystems (Buesseler et al., 2007), making it important for global models to predict changes in species composition caused by climate and other forms of global change. Furthermore, changing ocean stratification and thermohaline circulation, reduced extent of sea ice, and altered cloud-forming sulfate aerosols can profoundly influence ecosystem structure and function. Ocean acidification represents an emerging threat to the health of ocean ecosystems and its effects have only begun to be examined. Models need to be developed to study the potential biogeochemical, economic, and even sociopolitical impacts of ocean acidification in response to different emissions scenarios, similar to what has been done for the physical climate. Nonlinear feedbacks and thresholds are critical to understanding the complex responses of ecosystems and their future role in the carbon cycle.

New research is also needed to understand the impacts of rising atmospheric CO_2 and climate change on terrestrial ecosystems. Higher atmospheric CO_2 levels are likely to change the competitive balance among species, as seen in research showing that weedy species preferentially benefit at high CO_2 levels, thereby altering biodiversity. Another research priority is to assess how regional disturbances will change in the future. For instance, a southwestern United States that is warmer and drier is likely to experience increased fire frequency and severity, and may be more prone to outbreaks of insects, such as the mountain pine beetle. Large changes are also anticipated for arctic ecosystems, where research on primary production and permafrost vulnerability is vital.

More extensive study and enhanced measurements of marine and terrestrial ecosystem changes should be a key element of a comprehensive carbon cycle science strategy. Moreover, because sustaining healthy and diverse ecosystems is an important means of reducing greenhouse gas emissions in the face of changing climatic conditions (Turner et al., 2009), carbon cycle science must address strategies for preserving critical ecosystems and associated biodiversity.

4.4.4   Scientific directions

A scientific approach to address ecosystem impacts must involve a three-tiered effort that would 1) reduce uncertainties in understanding of, and ability to predict, ecosystem

responses to changes in greenhouse gas levels, 2) examine the synergistic effects of changes in atmospheric chemistry with changes in climate and other environmental parameters, and 3) sustain and enhance capabilities to observe changes in ecosystems as they occur.

Improve understanding of, and ability to predict, responses of ecosystem productivity, biodiversity, and sustainability to changing levels of carbon dioxide and other greenhouse gases.

Efforts to reduce uncertainties in our understanding of ecosystem impacts will require improved models supported by *in situ* and remote-sensing observations, as well as experimental manipulations and process studies that address changes in ecosystem productivity, biodiversity, and susceptibility to changing levels of CO_2 and other greenhouse gases. Studies should examine the effects of rising CO_2 as well as other greenhouse gases on terrestrial ecosystems and possible responses in productivity and community composition. Additional work should examine ocean ecosystem responses to multiple stressors including the effects of rising CO_2 and other gases, and their associated consequences. These efforts should also include work to examine ecosystem consequences of carbon sequestration strategies.

Determine the synergistic effects of rising CO_2 on ecosystems in the presence of altered patterns of climate and associated changes in weather, hydrology, sea level, and ocean circulation.

Additional efforts will be needed to determine the combined effects of rising CO_2 and altered patterns of climate on ecosystem structure and function in terrestrial and marine habitats. Although this broad topic extends well beyond the scope of this Plan and stimulates collaborative opportunities with other disciplines, aspects of this question fit well within the Plan's purview. Linkages between land and ocean ecosystems represent an issue that is particularly sensitive to change and that has important significance both for species and for society. These linkages are also only beginning to be examined in the context of carbon export to the coastal oceans and the impact of this export on coastal ocean acidification. Disproportionately large changes are also anticipated for Arctic and Antarctic ecosystems; consequently, a comprehensive science plan should include efforts to characterize ecosystem impacts in these regions and the ways in which they feed back to the carbon cycle. Interactions between human society and ecosystems must also be addressed as human activities have the potential to profoundly alter ecosystems on land and in the water. Of particular importance on land are ecological and climatic changes in the tropics, where food production and vulnerability to climate change are key concerns. Interactions between rising CO_2 and drought have been suggested in experimental studies.

Some key potential changes that would impact ecosystems and carbon feedbacks include changes in precipitation, soil moisture, and surface temperature with altitude and their impacts on natural vegetation and the distribution of managed agriculture; the susceptibility of species and ecosystems to disease, pests and fire; changes in precipitation and streamflow; changes in precipitation and surface temperatures (specifically at high latitudes affecting the spatial extent of permafrost); sea level rise impacts on coastal estuary, marsh, and ocean shelf ecosystems; and changes in ocean temperature, salinity, and circulation and related impacts on ocean ecosystems (e.g., coral reefs and primary productivity). Such areas of research require new focused integrative efforts bringing together multidisciplinary teams of researchers with expertise across physical climate, biogeochemical, and ecosystem sciences.

Enhance capabilities for sustained and integrated observations of ecosystems in support of scientific research as well as management and decision making.

Although targeted science goals to reduce uncertainties are important, immediate action is needed to develop our capabilities to observe ecosystems and provide critical information for scientific research as well as for environmental managers and decision makers. A comprehensive and integrated system of observations is essential for providing a baseline of existing conditions and the critical information needed to track and manage future change. Monitoring terrestrial and marine ecosystems is also a key component necessary for validating and refining models and identifying nonlinear responses and feedbacks.
Observational infrastructure should include terrestrial and ocean observation platforms, as well as remote-sensing observations, to provide time series of environmental conditions and ecosystem properties. Remote-sensing capabilities must be maintained and enhanced to enable larger scale tracking of changes in critical ecosystems. Additional technologies, including airborne sensors, unmanned

aeronautical vehicles, long-term field stations, moorings, floats, and underwater vehicles, can be used to further expand observational capabilities. These efforts should be integrated where possible with process studies to examine *in situ* responses to changing greenhouse gas concentrations and climatic forcing.

4.4.5   *Related Issues*

Because ecosystems play a fundamental role in mediating atmospheric levels of greenhouse gases, this goal is related to numerous other aspects of this Plan and to many aspects of ecology generally. Potential effects on the productivity and health of both marine and terrestrial ecosystems suggest the possibility of significant direct impacts on economically important species. Furthermore, an understanding of ecosystem dynamics is needed to develop accurate predictions of future changes and potential feedbacks and nonlinear responses. Finally, the impacts of increasing greenhouse gases on ecosystem structure and function are inextricably linked both to the capacity of these systems to sequester carbon, and to impacts from other elements of climate change, including links to hydrology, land use change, and sea level rise. Whereas impacts on carbon fluxes and storage are covered in other goals of this Plan, the broader set of feedbacks with climate change extend well beyond the scope of this Plan. Clear collaborations with, and linkages to, other scientific areas within the purview of the U.S. Global Change Research Program must be reinforced to coordinate research in this critical area.

4.5   Goal 5: Determine the likelihood of 'success' and the potential for side effects of carbon management pathways that might be undertaken to achieve a low-carbon future.

As concerns increase over anthropogenic impacts on the atmospheric concentrations of greenhouse gases and their impacts on the global carbon cycle, it is critically important to determine the likelihood of success and the potential for undesirable side effects of possible carbon management pathways to achieve a low-carbon future. This goal aims to understand interlinked natural and managed systems sufficiently for individuals, corporations, and governments to make rational and well-informed decisions on how best to manage the global carbon cycle, and especially the anthropogenic impacts on this cycle.

4.5.1 Motivation

The global carbon cycle is complex and closely linked to Earth's energy, water, and nutrient cycles and to demographic and economic systems globally. Efforts to manage the carbon cycle will have broad environmental and economic impacts. Ethical and equity issues are central to what actions might be taken, who takes them, and what consequences result. The many interconnected factors affected by carbon management strategies must be understood and taken into account to determine the likelihood of success of alternative carbon management schemes.

In addition, low-carbon strategies have the potential to harm local and/or distant ecosystems and communities. Issues characterized as 'food vs. fuel' or 'indirect land use change' represent the emerging concerns about the impacts of carbon mitigation strategies. Those systems that utilize large land areas could potentially displace small landholders and sharecroppers through land consolidation to produce biomass or to harvest solar or wind energy; they will similarly impact the Earth's surface energy balance, biodiversity, and water balance. Proposals to inject CO_2 into the deep ocean have been diverted by concerns about effects on marine ecosystems, and carbon sequestration in the biosphere has raised questions about changes in albedo. All of these interconnections among environmental and economic concerns require that we have a clear understanding of the impacts of alternatives, including both the aggregate impacts on the global system and the distribution of these impacts regionally and locally.

4.5.2 Progress over the last decade

Considerable progress has been made over the last decade in determining the net greenhouse gas balance of some carbon mitigation activities. For instance, current corn-ethanol production technology has been shown to have a less positive carbon balance than originally estimated by some scientists and policymakers (e.g., Fargione et al., 2008; Searchinger et al., 2008; Piñeiro et al., 2009). As such, the magnitude of the climate benefits of the 2007 Energy Independence and Security Act mandating the use of 36 billion gallons of biofuel by 2022 have been questioned. Clear estimates of carbon savings along with the potential consequences of carbon sequestration strategies for other greenhouse gases, such as methane and nitrous oxide, water, and other ecosystem services are urgently needed (e.g., Jackson et al., 2005). Carbon

capture and storage has undergone considerable technical evaluation and attracted much interest, but additional issues such as monitoring for possible leakage have a clear connection to carbon cycle research. Many potential decisions have both positive and negative consequences and scientists need to provide a comprehensive, quantitative analysis of the tradeoffs among greenhouse gas emissions, other environmental impacts, economic and social impacts, and the distribution of costs and benefits.

4.5.3 Major uncertainties

Low-carbon futures will impact both environmental and economic systems and we are just beginning to understand the range and magnitude of the issues. Questions as apparently simple as the tradeoffs between the capital investment in a new car and the savings that will be achieved during operation are important to confront, in both environmental and economic terms. As low-carbon strategies are implemented, both the environmental and social impacts will be felt in different places and sectors of the communities undertaking the effort – changes in energy availability and cost; access to various resources such as land, water, and food production; and livelihoods will be apparent. Not all segments of society will equally share in the profits and the burden from the shift of employment, economic gains, and environmental improvement, and these differential costs and benefits are poorly understood. In many cases we cannot accurately characterize the aggregate costs and benefits, let alone the distribution of those costs and benefits. For low-carbon strategies that achieve aggregate climate and other environmental benefits, institutional structures will be needed to motivate adoption and to provide oversight in the sharing of gains and losses as deployment of the low-carbon strategies are carried out. Additional research is needed into what institutional structures will be most effective in providing motivation, oversight, and verification of carbon management goals. The complexities for interdisciplinary collaboration and international cooperation are myriad and require explicit attention.

4.5.4 Scientific directions

As society moves into a phase of active carbon management, several research issues need to be addressed to determine the efficacy of proposed strategies.

Develop mechanisms for evaluating and integrating interconnected and potentially competing management goals within the context of carbon cycle science

Continuing the biofuels example from above, in the development of a low-carbon strategy there are many considerations associated with societal goals. These considerations include climate protection, food security, human livelihoods and well-being, conservation of biodiversity, and maintenance of ecosystem services of land and ocean domains, as examples. Recent concerns have been raised over the potential impacts of bioenergy on socio-environmental dimensions. The challenge for carbon cycle science is to provide quantitative and balanced observations, data, and analyses that the policymaking community can use to balance societal objectives in an informed way. A critical initial step is to ensure that strategies to reduce greenhouse gas emissions do indeed reduce net greenhouse gas emissions. Whether these are technical strategies like development of cellulosic ethanol or social strategies like promotion of mass transit, detailed systems analyses are needed to ensure that greenhouse gas benefits are real. The challenge in devising any low-carbon development strategy is to collectively evaluate multiple goals while recognizing the tradeoffs across the suite of benefits and liabilities that comprise the decision process. Research is needed to identify mechanisms that minimize the net negative effects across a spectrum of goals. An integrated approach to evaluating the impact of these low-carbon strategies is also needed to better understand the impact these strategies have on the environment and on socioeconomic factors.

Determine the impacts of carbon management and sequestration strategies on sustainability of ecosystems and ecosystem services, including water resources and biodiversity.

Sustaining healthy and diverse ecosystems is an important means of minimizing greenhouse emissions and maintaining ecosystem resiliency in the face of the changing carbon cycle and climate (Turner et al., 2009). Carbon cycle science must address current uncertainties about land and marine management opportunities to sequester carbon while preserving biodiversity, water resources, and other critical ecosystem functions. Carbon cycle science needs to collaborate with other science disciplines to provide the knowledge for informed decisions.

Carbon management strategies pursued in the biosphere will have multiple impacts on the climate system due to their impact on Earth surface properties such as reflectivity, evaporation, and surface roughness. It is important for carbon cycle scientists to understand the carbon cycle science with sufficient precision and to collaborate with climate scientists with sufficient intensity to ensure that activities undertaken with the goal of mitigating climate change do indeed accomplish the primary goal. Important research questions remain about the nature, as well as the temporal and spatial dimensions, of carbon management.

4.5.5   Related Issues

Understanding the impacts and benefits of carbon management policies will require collaborations far beyond the traditional boundaries of the carbon cycle science research community. Research communities investigating the human dimensions of climate change and carbon policies need to be working together with carbon cycle scientists. This goal, as well as much of our new Plan, calls for unprecedented levels of integration and cooperation among multiple research communities.

4.6   Goal 6: Address decision maker needs for current and future carbon cycle information and provide data and projections that are relevant, credible, and legitimate for their decisions.

One purpose of this research program is to support decision making at many different scales as society responds to the challenge of climate change. This sixth goal seeks to understand how decision making affects the evolution of the carbon cycle, determine how information about the carbon cycle can be relevant to policy decisions and ultimately to provide carbon cycle information needed by decision makers. As used here, the term 'decision makers' is meant in its broadest sense to include the general public, stakeholders, policymakers, and many other groups. Goal 6 addresses the research agenda that is necessary for carbon cycle science to more effectively support decision making. This is distinguished from the infrastructure development and other activities that are needed to support communication efforts from the program more generally, as are described in Section 5.4. This goal recognizes the need to be anticipatory. The needs of decision makers a decade from now will not necessarily be the same as their current needs, and a goal of research needs to be to anticipate and probe so that we are prepared to confront tomorrow's questions.

4.6.1 Motivation

Sound, understandable carbon science is needed by a broad range of decision makers, including government and business leaders, as they formulate strategies to address climate change. As carbon policies are advancing rapidly in both the private and public sectors, carbon science has the potential to inform decision making at many scales. However, research in the field of climate-related decision support suggests that an interactive, deliberate approach must be used to understand what is relevant and useful to these decision makers (e.g., Lemos and Morehouse, 2005; NRC, 2009a; Dilling and Lemos, 2011). Several principles have emerged for tackling research in this area, starting with orienting a decision support program around users' needs for information (NRC, 2009a). Moreover, understanding the contexts for decisions that affect the carbon balance is critical for being able to project the development of the carbon balance in the next century (see Goals 1 and 2). We therefore propose an invigorated, interdisciplinary effort that will understand these decision-making contexts in order to provide science that is usable to decision makers.

4.6.2 Progress over the last decade

Formal links between the carbon science community and policymakers at the international level are well established through the IPCC process and through reports on topics such as carbon capture and storage, inventory methods, land use change and forestry, and emissions scenarios (e.g., IPCC, 2006; Dilling, 2007; see IPCC, 2011). The goal of supporting policy and decision making is echoed in many scientific organizations including the Global Carbon Project and the International Council for Science (ICSU). In the United States, carbon cycle science is also positioned to be relevant for decision making at national, state, and local levels through individual projects as well as agency-wide efforts (e.g., the U.S. Forest Service; Hudiburg et al., 2009; Potter, 2010; Law and Harmon, 2011). However, explicit understanding is needed of the types of carbon cycle science that are required for decision making, how carbon information relates to other variables of

interest, and how to create an iterative, ongoing capacity to connect researchers in this area to the user community. Issues of current focus such as the development of renewable energy, emissions trading, and preserving stocks of carbon in forests and soils will rely on understanding the global carbon cycle. Finally, the program needs to develop metrics by which to judge whether decision support activities are effective (e.g., Moser, 2009).

4.6.3 Major uncertainties

Given the societal importance of decarbonizing the economy and preserving terrestrial carbon stores, there are many decision-making contexts that could be informed by carbon science at appropriate scales. Opportunities exist to build ongoing relationships with particular sectors and stakeholders in order to understand what carbon science is relevant and useful. One major uncertainty that exists outside of the realm of the research enterprise is whether, and how, mechanisms might be set up that effectively constrain carbon emissions. Because carbon emissions are still an externality that is not valued explicitly in most nations' policy structures, decision makers may have quite different needs and demands for carbon information. The demand for reliable carbon information is likely to be quite high if mandatory carbon reduction policies are put into place, but may be lower if carbon remains a voluntary consideration for decision makers (Logar and Conant, 2007). The policy context for this research is therefore quite uncertain. Political, economic, and other forces are likely to constrain how important carbon information is with respect to other tradeoffs within the decision space. A second type of uncertainty involves the impact of various decisions themselves on the carbon balance. For example, forestry management in the United States currently places a high priority on reducing fuel loads in order to prevent catastrophic level forest fires from occurring near urban settlements (e.g., Kashian et al., 2006). How forest management decisions affect the carbon balance is still an active area of research. A corollary to this argument suggests that there is a need to understand how to present carbon information in the context of the decision and tradeoffs to be made so that it can be of most use to decisions in the field.

4.6.4   Scientific directions

The NRC (2009a) divides research for decision support into two broad categories, research *for* decision support, and research *on* decision support. Many of the goals listed above in this plan fall into the category of research *for* decision support: 1) understanding vulnerabilities associated with changes in the carbon cycle and human development scenarios; 2) understanding the potential for mitigation; 3) understanding adaptation contexts and capacities; 4) understanding how adaptation and mitigation interact with each other; and 5) understanding and taking advantage of emerging opportunities associated with climate variability and change (NRC 2009a). This goal also includes research *on* decision support, namely: 1) understanding information needs; 2) characterizing and understanding risk and uncertainty; 3) understanding and improving processes related to decision support; 4) developing and disseminating decision support products (see also Section 5.4); and 5) assessing decision support experiments (NRC, 2009a). We list here two areas of scientific direction, the first related to research *for* decision support, and the second related to research *on* decision support.

Characterize the fundamental dynamics of decision making as they affect large-scale trends and patterns in carbon stocks and flows

Identifying key processes and drivers that control carbon fluxes is central to studying carbon decision-making dynamics. In land use decisions, for instance, drivers such as values, climate, global markets, economic pressures, opportunities, and policies at various scales are all important to decision makers (Richards et al., 2006). Decisions at various scales can and do intersect, and collectively they often result in emergent patterns of carbon storage or fluxes, depending on the interactions. In addition, a given policy direction can have unexpected results depending on the situation and receptivity of individual decision makers to the policy, as well as interactions with global markets. In all likelihood, carbon management as a goal for decision makers will continue to be embedded within the context of multiple, sometimes competing goals (Tschakert et al., 2008; Failey and Dilling, 2010). Research here will build on efforts of social science programs such as those in land use change (e.g., Global Land Project), urbanization and carbon (e.g., Urbanization and Global Environmental Change Project), and integrated assessment modeling to conduct

interdisciplinary research that connects decision making (typically studied through social science methods) to carbon outcomes (typically studied by natural science methods).

Systematically address decision maker needs for carbon cycle science information as they begin to incorporate carbon-related factors into their decision making

Several models exist of communities that have successfully devised approaches for making science 'usable' for decision making. One example in a closely related field is the seasonal to interannual weather forecasting community. Overall, usable science is science that is relevant to the decision context, that is available at the time and geographical scale of interest, and that can make a difference in the outcome of a decision (Lemos and Morehouse, 2005; Tschakert et al., 2008; Dilling and Lemos, 2011). The four main requirements in making science useful for decision making are: 1) to understand the context in which information is to be provided; 2) to ensure the information produced is relevant to the decision and that realistic choices are available; 3) to confirm that a receptive institutional, cultural, and organizational setting exists into which information can enter; and 4) to establish that adequate information and delivery mechanisms are present. To create these conditions, an ongoing, two-way dialogue between researchers and decision makers must be established early and maintained over time, in order to build trust and knowledge of what is possible scientifically, and of what is useful from a decision-making perspective (Morss et al., 2005; Lowrey et al., 2009).

An example of this dynamic would be trying to understand the issues surrounding geoengineering approaches to mitigate climate change. In this context, the public (who may be asked to pay for these approaches) could have one set of motivations, companies interested in implementing geoengineering approaches might have a different set of motivations, and the policymakers may have yet a different set of needs and motivations. Decision making relevant to carbon outcomes is pervasive, and the challenge will be to identify those potential users who will benefit from direct linkages with the carbon cycle science community and who represent the best fit with the existing mandate of the carbon community for providing decision support. A series of pilot projects at various scales will be necessary to develop skills in this area so that a broader effort at improving decision making can occur over time.

4.6.5 Related Issues

Research related to decision making for carbon has been carried out beyond the carbon cycle science community, in arenas such as economics, integrated assessment modeling, land use and land cover change, and even energy technology. These communities are beginning to intersect with the carbon community through international efforts in the International Human Dimensions Programme on Global Change through the Earth System Science Partnership to form research agendas, joint projects, and shared questions, but the effort is still maturing. Integrating the knowledge and background of people from communities outside of the traditional carbon science arena will be important for the success of Goal 6, including developing new areas of interdisciplinary research on decision making and the carbon cycle.

Chapter 5
Science Plan Elements

A number of key research components comprise the central core for advancing carbon cycle science over the next decade. We group these cross-cutting components into four high priority elements: (1) sustained observations; (2) studies of system dynamics and function across scales; (3) modeling, prediction, and synthesis; and (4) communication and dissemination. These are the 'action items' of the research agenda. Each of these elements contributes to all six of the goals described in the previous chapter, and all four are critical to achieving each of the goals. In the text that follows, these elements are described to provide finer focus, including details and examples of the types of research that are needed and how current research activities need to be enhanced to fully realize the goals stated in Chapter 4. The most critical priorities for each of these elements are highlighted in *italics*.

5.1 Sustained observations

The observational network for measuring and tracking carbon/climate is the backbone of the global carbon program. The measurement programs document the evolution of the carbon cycle in the atmosphere, terrestrial biosphere, and ocean, as well as the human systems that affect these carbon reservoirs. The key to an effective carbon/climate observational network is continuity of measurements, adequate spatial and temporal coverage, and the development of long-term records. These observations must be integrated into an observational 'network of networks' to ensure that scientific data and insights are leveraged across observations platforms and across components of the carbon system.

Key components of such an observation system include: satellite observations of Earth surface properties and atmospheric constituents; atmospheric observation networks (including flux, flask, tower, and airborne observations); open-ocean as well as coastal-ocean surface and subsurface sampling (including organic and inorganic carbon, relevant tracers, and biological observations); systematic measurements of terrestrial biomass, carbon fluxes, and biodiversity; and monitoring and assessment of human systems – including mitigation and adaptation strategies and associated impacts. All of these observations have an inherent time scale that has the greatest relevance for carbon cycle studies. For example, while some observations require near continuous measurements to fully document the variability, others may require only thorough surveys every few years or once per decade. We do not prescribe here what those time scales are or attempt to detail exhaustively the full breadth of the components described below.

One issue common to all of these observation components is that data are typically gathered through research projects with grant-driven funding cycles and spread across multiple agencies. While the research focus helps to ensure that the measurements are state-of-the-art and relevant to key research questions, it also means that these observations generally have uncertain long-term funding and limited coordination.

More stable funding options must be identified for the subset of these observations that forms the backbone of our understanding and study of long-term carbon cycling dynamics. Strategies for ensuring data continuity and the funding mechanisms to make this possible will need to also take into account recent interest by private entities to fund, build, and maintain observation networks as potential profit-making activities.

5.1.1 Atmospheric observations

Observations that provide data on atmospheric abundances of CO_2, CH_4, and related species, as well as plot-scale observations of carbon fluxes, are providing key long-term data for carbon cycle science.

Existing *in situ* networks will benefit from a more coordinated and integrated design, together with longer-term sustained funding for the key elements. For example, global carbon flux networks such as FLUXNET, and its American subset, AmeriFlux (ORNL, 2010), provide information on carbon sources and sinks in different terrestrial ecosystems. These flux towers use observations of atmospheric abundance of CO_2, together with measurements of turbulence statistics, to estimate carbon fluxes within the footprint (typically approximately 1 km^2) of the towers. However, the individual sites in these networks are typically funded on a regular grant cycle, and many sites are therefore being discontinued due to a lack of funding availability. The atmospheric CO_2 flask and continuous-observation tower network, coordinated by NOAA ESRL, is a cooperative network documenting trends in atmospheric CO_2 concentrations that give insights into carbon flux at large spatial and temporal scales, but this network too has issues of site continuity. NOAA ESRL also maintains regular aircraft-based atmospheric sampling at some sites, supplemented by additional data provided through investigator-led research projects.

A high-priority need in this area is to select and standardize a subset of flux, flask, tower, ship, buoy, and aircraft sites coordinated as a permanent network with steady funding to ensure data continuity. The NSF-funded National Ecological Observatory Network (NEON) project (NEON, 2010) will fill a part of this need from the standpoint of flux observations over the next five years as it constructs 19 bio-climatologically defined domains that will each have one to three flux towers. The NEON network should complement the existing AmeriFlux network, including the more than 20 current AmeriFlux sites that have been in operation for at least a decade. A coordinated network should encompass not just existing observation sites but should expand in some key components. Additional tall towers that provide continuous CO_2 observations are needed to constrain the North American carbon cycle. Aircraft profiles and large-scale transects of atmospheric sampling also provide valuable regional scaling,

and regular routing and time schedules would enhance the information that they provide.

In addition, although much of the emphasis has been on expanding and maintaining CO_2 observing capabilities, measurements of several other atmospheric species (including, but not limited to CH_4, CO, and carbon isotopes) also form a critical component of a stable observing network.

5.1.2 Ocean, coastal, and inland water observations

Surface ocean carbon observations are currently made on research ships, volunteer observing ships (VOS), moorings, surface drifters, and from satellites. The current *in situ* network is relatively strong in the North Atlantic and North Pacific but less so for other ocean basins. Additional systems are needed in a number of key regions. Interior ocean carbon observations have made good progress in documenting changes in ocean physics, carbon, and other tracers since the World Ocean Circulation Experiment/Joint Global Ocean Flux Study (WOCE/JGOFS) cruises of the 1990s (see NOAA, 2010), but needs to be maintained to understand ongoing changes. These observational networks are reasonably well coordinated but require a more stable long-term funding structure and ship time to help ensure their continuity and to build out the networks in under-sampled regions.

Although there are a number of measurement programs in coastal waters (e.g., NSF Long Term Ecological Research (LTER) sites, NSF Ocean Observatories Initiative (OOI), NOAA Integrated Ocean Observing System (IOOS)), very few have a carbon focus and there is almost no large-scale coordination. A coordinated biogeochemical observing network for U.S. coastal waters would provide consistency in what is measured, the frequency of observations, and the reporting of data to national data centers. A U.S. coastal observing program could build on existing infrastructure to coordinate observations that would not only continue to serve the local needs but also contribute to a large-scale coastal carbon observational effort.

Stronger connections are also needed between coastal ocean studies, where terrestrial carbon is considered an input, and land carbon studies, where carbon exported to rivers and the ocean is considered a loss term. The lateral exchanges of carbon between the continuum of land, fresh water systems, and the ocean need to be studied together in addition to the vertical exchanges with the atmosphere. *Overall, there is a high priority need to expand and enhance the open-ocean and coastal carbon observational networks including stronger ties to the terrestrial carbon program.*

Furthermore, there is a high priority need to establish a measurement and observational effort aimed at understanding biological processes and ocean acidification. Primary production, carbon and nitrogen fixation, metabolism, and biological species composition are important biological measurements for understanding the ocean carbon cycle. Coordinated global-scale biological observations can provide insight to marine population- and community-level changes and could ultimately lead to development of biological indicators. Of particular importance is development of indicators that characterize the biological effects of ocean acidification.

The existing ocean interior observations do provide geochemical data relevant to large-scale ocean acidification but the current surface observing network is geared primarily toward quantifying air-sea gas exchange so it focuses on CO_2 partial pressure measurements. To constrain the observations of changes associated with ocean acidification, the network needs to be enhanced to add a second carbon parameter and supporting biological observations in the surface ocean. The coastal network described above should also provide the information needed on ocean acidification in the coastal regions. Special attention should be focused on observing and tracking the process and impacts of ocean acidification in particularly vulnerable environments such as coral reefs. Ecosystem studies need to be comprehensive and recognize that organisms are facing multiple stressors. In some cases these stressors may have interactive effects (e.g., increasing CO_2 and temperature together can have a larger impact on an organism than changing either parameter individually).

5.1.3   *Terrestrial ecosystem observations*

In situ terrestrial observations are currently made at a range of sites and using differing technologies. For instance, the NSF LTER sites provide sustained observations at 26 locations, primarily in the continental United States. As discussed briefly in Section 5.1.1, the NSF-funded NEON project will enhance terrestrial observations of carbon pools and fluxes over the next years as it constructs measurement centers in 19 regions of the

United States. There are currently more than 80 AmeriFlux sites operating in different biomes and in clusters along disturbance/climate gradients across the United States. The use of this observation network, also described in Section 5.1.1, includes estimates of water and energy fluxes, meteorological data, and information on biological processes to understand responses of component fluxes and the integrated total ecosystem responses. More than 20 sites have been running for at least 10 years, making it possible to analyze trends in carbon fluxes with succession, disturbance processes of insect attack, fire, harvest, and changing hydrology. This existing network has a North American focus; coordination with other terrestrial observing networks is needed for integrated carbon cycle research.

The best current terrestrial inventory system in the United States is the Forest Inventory and Analysis (FIA) program, which evolved from the permanent forest growth plot data of the US Forest Service (USDA Forest Service, 2010). The long-term FIA data can be made even more valuable by improving public access for researchers to higher spatial resolution observations and by developing standardized gridded products for use by the broader carbon cycle modeling community. Using the FIA model as an example, an equivalent regular inventory of agricultural and rangeland soil and ecosystem carbon should be developed to provide a more comprehensive view of land-based carbon stocks. Together with the flux tower observations, the measurements would provide critical understanding of the terrestrial component of the carbon cycle in North America and globally. Such observations should be closely coordinated with related international efforts to maximize the impact of the overall observational effort. These and other observing networks will also allow terrestrial carbon research to address key feedbacks with climate and with water, nitrogen, and other interacting variables that are important for carbon cycling but are outside the scope of this plan. Examples include CO_2 fertilization under the constraints of nitrogen mineralization and deposition, the vulnerability of tropical forests in the face of droughts and other decadal trends in precipitation, the balance between enhanced decomposition rates and rising gross primary productivity (GPP) due to enhanced N-mineralization in temperate and boreal ecosystems with soil warming, and the destabilization of vast amounts of organic matter because of melting permafrost, leading to large releases of atmospheric CO_2 and CH_4.

Overall, there is a high priority need to establish and standardize regular cross-agency observations of terrestrial carbon variables for forests, agricultural lands, and rangelands, and to establish a more comprehensive land- and satellite-based network to assess land use effects on the carbon cycle. Maintaining continuity and coordination for the terrestrial observing network, and its integration with remotely sensed observations, is critical.

5.1.4   *Monitoring and assessment of human systems, including mitigation and adaptation strategies and associated impacts*

Sustained data collection and monitoring of human activities resulting in CO_2 and CH_4 emissions are critical for understanding the patterns of CO_2 uptake and release. A broad range of demographic, economic, and technologic data are needed for understanding, projecting, and potentially managing the human role in the global carbon cycle.

Information on fuel use for electricity generation, transportation technology and practice, fossil fuel use for home heating, energy resources, and trade are just a few examples of the types of information that are needed in the energy sector. Some of this information is already being collected by national agencies and in international compilations but regional and national data are of inconsistent character and quality. Data are also needed at finer spatial and temporal scales than countries and years. The energy sector is an example of the importance of global collaboration as the roles of international markets and international trade affect the global energy system. International data on renewable energy are notoriously weak, and the development of biomass energy systems interacts closely with carbon stocks and flows in the terrestrial biosphere.

Data on large-scale land use, including farming, forest management, urban development, wetland distribution, and coastal zone modification need to be collected regularly. But gross patterns of land use do not reveal the changes and motivations that can be gleaned from patterns of land ownership, land productivity, land fragmentation, and management practice. Data on urban development and policy development are needed to interpret how emissions from urban centers vary in space and are evolving over time. Census data and survey information on topics such as attitudes toward climate change, management policies, and fossil fuel

regulations will provide valuable information on the drivers of social decisions and their evolution over time.

Economic data need to be collected on carbon commodities (such as wood products), materials substitution, waste management, and industrial energy demand. Additional data are needed on industries that will be directly or indirectly affected by carbon management policies, changes in the carbon cycle, or changes in the climate system. These data will provide information on the evolution of human carbon system drivers as well as feed into the cost/benefit analyses needed to evaluate management pathways.

Synthesis and attribution projects that rely on a steady stream of information collected at local, regional, and national levels will be critical for monitoring the response to carbon/climate change and mitigation efforts. Separation of the human and 'natural' components of the carbon cycle depends on understanding controlling processes over a wide range of temporal and spatial scales.

Many local, state, and regional mitigation and adaptation strategies are being developed to reduce carbon footprints and respond to climate change. As these strategies are implemented, they need to be monitored to determine their effectiveness and to identify any potential biogeochemical or social side effects. A large-scale data collection effort can be used to inform small-scale mitigation and adaptation projects focused on identifying the strategies that are most effective and cost efficient.

Of the sustained observations sub-elements, this priority on human systems is the least developed in the scientific community and requires substantial new resources to develop and coordinate and to design the most efficient and effective observational systems. Funds should prioritize integrated social and natural science projects where possible.

5.1.5   *Remote-sensing observations of the Earth system*

Satellite and other remote-sensing observations of the Earth system complement the *in situ* observations described in the previous sections. Earth-observing satellites provide global data at high spatial and temporal resolutions, making it possible to evaluate global patterns in the carbon system in a manner that would not be feasible using only *in situ* observations.

On land, remote-sensing observations of ecosystem structure, carbon stocks, disturbance, land use and land cover change, and phenology provide key insights into global trends. In the oceans, remote sensing is critical for understanding global patterns of ocean physics (e.g., temperature, dynamic height), biology (e.g., ocean color), chemistry (e.g., salinity), and air-sea forcing properties (e.g., surface winds, wave height). Satellite observations of atmospheric composition, including CO_2 and CH_4, are complemented by observations of ancillary data that inform carbon processes, such as cloud and aerosol properties, precipitation, and temperature. A comprehensive list and description of current and upcoming Earth observing satellites that will contribute to carbon cycle science is beyond the scope of this Plan. Examples are provided here, and a recent review of mission needs and capabilities is available as part of the Group on Earth Observations (GEO) Carbon Strategy report (Ciais et al., 2010).

Satellites from the NASA Earth Observing System (EOS) platforms provide key atmospheric, terrestrial, and oceanic observations that have played a central role over the last decade, and will continue to do so for the lifetime of these missions. Examples include the documentation of land use change using the series of Landsat satellites, and the innumerable studies made possible by the phonological observations provided by the Moderate Resolution Imaging Spectroradiometer (MODIS) instruments on Terra and Aqua. EOS satellites also provide observations of Earth surface properties necessary for assessing carbon stocks (e.g., Landsat-7, Terra, Aqua) and of atmospheric composition including the mid-tropospheric CO_2 observations provided by the Atmospheric Infrared Sounder (AIRS) on Aqua and the Atmospheric Infrared Sounder (TES) on Aura. As we enter a new decade, however, many satellites from the EOS platforms are nearing the end of their operational lifetimes.

Given the need for data continuity, making the transition to the Joint Polar Satellite System (JPSS) operational platform, and the preparatory National Polar-orbiting Operational Environmental Satellite System (NPOESS) Preparatory Project (NPP) mission, will be a critical activity in the coming decade. Algorithms and data products from current instruments will need to be replicated using the new generation of instruments. In addition, data processing systems will need to be developed for JPSS that are compatible with existing datasets.

Several other missions currently in formulation and implementation will also provide critical data for carbon cycle science in the coming decade. Examples on land include the Landsat Data Continuity Mission (LDCM) and the Ice, Cloud, and land Elevation Satellite-2 (ICESAT-2), which will help to assess carbon stocks, and Soil Moisture Active and Passive (SMAP), which will inform linkages between the water and carbon cycles. In the atmosphere, OCO-2, a replacement for the OCO satellite lost during launch in 2009, will become the first NASA mission designed specifically for making CO_2 observations from space. In the oceans, the Visible Infrared Imager Radiometer Suite (VIIRS) instrument aboard NPP and JPSS will provide the next generation of observations of sea surface temperature and ocean color.

In addition, the National Research Council (NRC) Decadal Survey (NRC, 2007a) recommended the development of several satellites that will provide high-value observations for carbon cycle science in the next decade or shortly thereafter. The ICESAT-2 and SMAP satellites are part of this survey. Satellites being planned for later launch dates will further elucidate carbon cycle processes on land, in the atmosphere, and in the oceans. Examples for terrestrial ecosystems include the Hyperspectral Infrared Imager (HyspIRI), which will monitor vegetation type and function to detect responses of ecosystems to human land management and climate change and variability, and Lidar Surface Topography (LIST), which will inform global shifts in vegetation patterns and forest stand structure by mapping global topography at high resolution. The ASCENDS satellite will provide the next generation of satellite-based CO_2 atmospheric observations, and was recently recommended for accelerated development. In aquatic systems, the Aerosol–Cloud–Ecosystems (ACE) and Pre-ACE (PACE) satellites will inform ocean biogeochemistry to understand the evolution of ocean ecosystems in response to ocean acidification and climate change, while Geostationary Coastal and Air Pollution Events (GEO-CAPE) will study coastal carbon processes.

Collaborations with international partners will make it possible to further expand the science both on land (e.g., Phased Array type L-band Synthetic Aperture Radar (PALSAR) on the Advanced Land Observing Satellite (ALOS)) and for the atmosphere (e.g., CO_2 and CH_4 observations from the Greenhouse Gas Observing Satellite (GOSAT), the

Infrared Atmospheric Sounding Interferometer (IASI), and the Scanning Imaging Absorption Spectrometer for Atmospheric Cartography (SCIAMACHY)).

In addition to space-based observations, aircraft-based remote sensing provides valuable observations at local to regional scales, while also potentially serving as a testbed for space-based instruments. Ground-based remote sensing, such as the Total Column Carbon Observing Network (TCCON) network, directly informs carbon cycle science, while simultaneously providing key validation data for space-based instruments. *The use of ground- and aircraft-based remote sensing should remain a key complement to in situ and space-based observations.*

A high-priority activity is to establish long-term continuity of critical satellite-based datasets of the Earth system critical to improving our understanding of the carbon cycle. Recent budget cuts are putting such continuity at risk, especially for observations of ecosystem structure and observations that allow coupling of the understanding of carbon with climate dynamics.

5.1.6 *Mapping sustained observations into the goals*

The observations that comprise this element contribute to all of the stated goals but are particularly important to Goal 1. One cannot have a clear and timely explanation of past and current variations observed in atmospheric CO_2 and CH_4 without knowing what those variations are or how the other reservoirs and drivers have changed over time. Observations are needed over a range of temporal and spatial scales to more effectively attribute observed changes to particular processes. Sustained observations are also needed for Goal 2. The human system observations are critical for understanding the socioeconomic drivers of emissions, and all components of this element are needed for monitoring and verifying emissions. Fossil fuel and land use observations coupled with natural science observations are needed to estimate emissions and to provide independent assessments of those emissions. Knowing the history and records of change of the pools will help provide an understanding of potential vulnerabilities as a component of Goal 3 and how ecosystems and species are changing for Goal 4. Observations of human activity will also provide information on how human systems are changing as a result of increasing awareness of vulnerabilities. Quantifying

the amount of carbon in plants and other organisms globally will also help to determine which components of ecosystems are most vulnerable to rising CO_2. Goal 5 requires sustained observations to document the results of what has happened under different management strategies, including how effective the strategy was and how humans have responded to the strategies. Finally, through Goal 6, decision makers will rely on observations to confirm that their decisions are having the desired effect.

5.2 Process studies of system dynamics and function across scales

Quantitative understanding of processes that affect carbon cycle dynamics across a spectrum of spatial and temporal scales is important for diagnosing and predicting how the carbon cycle responds to changes in fossil fuel use, carbon management policies, atmospheric composition, climate, nutrient availability, disturbance, land management, and other drivers. Our understanding of how the carbon dynamics of human, terrestrial, and ocean systems respond to, and interact with, changes in these drivers is incomplete. This is evidenced by the wide range of predictions of the future carbon balance of the terrestrial and ocean systems, as well as that of future anthropogenic carbon emissions. Understanding basic processes is particularly important in a system that is as complex, with interacting factors, as is the global carbon cycle.

Process studies are critical for achieving each of the six goals outlined in this Plan. These studies include efforts to provide the mechanistic understanding for improving diagnostic and prognostic models of the carbon cycle. Manipulative experiments are an important complement to observational process studies of the current state of the carbon cycle for two reasons. First, experimental studies extend process studies into environmental and socioeconomic conditions that may occur in the future, challenging our mechanistic understanding of how the carbon cycle will function in altered environments. Second, manipulative experiments and process studies provide complementary understanding for informing and parameterizing the response of predictive carbon cycle models to evolving environmental drivers. Process studies will alert us when changes in the carbon cycle, the climate, or their consequences imply either positive or negative consequences for natural or economic systems.

5.2.1 Intensive process studies and field campaigns

A major opportunity is to conduct intensive process studies across traditional disciplinary boundaries to observe and understand the natural and human systems and the processes controlling carbon emissions, uptake, and storage. Intensive campaigns make it possible to focus on a time, space, or environment in a way that limits some variables while focusing on others to facilitate analysis. Process studies can be designed to fill basic research gaps on land, in the oceans, or in transitional environments (including those systems with potential for large loss of stored carbon), and to integrate how socioeconomic issues and responses influence human impacts on carbon uptake and storage.

In terrestrial systems, the development of improved understanding of the carbon cycle can arise from sustained observations and from short-term but focused studies in ecosystems vulnerable to carbon loss, including permafrost, forests, boreal and tropical peatlands, and locations with methane hydrates. Intensive campaigns can provide observations at very fine spatial and temporal scales that are too expensive to pursue in long-term measurement networks. These observations should emphasize, for example, the carbon not just in plant biomass but in soils as well and should look at changes over small distances or time scales to provide mechanistic insight in the causes of those changes. Additional observations are needed to understand the processes responsible for human fossil fuel use, waste streams, land use, land management, and other factors, but here too, intensive studies can lead to improved process understanding. The impacts of human activities are linked to storage of carbon in the ocean and terrestrial ecosystems, and these are linked in complex ways to changes in atmospheric composition, nitrogen cycling, climate, disturbance regimes, land management, and factors affecting methane emissions. The behavior, functioning, and transport of river systems, for example, can be much influenced by how humans are impacted and respond to short-term extreme events.

Similarly, many basic aspects of the ocean carbon cycle are inadequately understood and would yield process understanding in intensive, focused campaigns. Important processes that need to be better quantified include gas exchange at the surface and the rate of anthropogenic carbon transport from the mixed layer into the thermocline and deep ocean. Also important is the role of biological processes in determining the spatial and temporal variability of air-sea fluxes and anthropogenic carbon uptake and storage. A deeper understanding of what controls the biological pump in the ocean is required, including the role of micronutrients and CO_2 in controlling productivity, controls on export of organic material from the surface, and transformations of organic material below the sunlit surface layer of the ocean. Detailed studies during specific events or in focused environments can provide notable insight.

Linking terrestrial and ocean systems is another important opportunity for better understanding of how carbon, nutrients, and sediments are moved from terrestrial ecosystems through estuaries to the ocean, where the fate of carbon can be long-term storage in the marine environment or release to the atmosphere as CO_2. Some environments are of limited geographic extent yet play major roles in the global carbon cycle and these provide rich possibilities for focused research. The transport of carbon into and through rivers and other freshwater networks, the transformations of these constituents in these networks, and the delivery and fate of this carbon in deltas and coastal ecosystems; including the processes that control the conversion and loss of carbon in coastal oceans and along continental margins often take place over limited regions and vary significantly with time.

5.2.2 Manipulative laboratory and field studies

Manipulative laboratory and field studies are important for elucidating the response of land and marine ecosystems to climate, biogeochemical, and socioeconomic change and to intentional carbon management. Much progress has been made in the last two decades in studying responses of terrestrial and ocean ecosystems to manipulations of individual geochemical or climatic factors. For example, studies examining the effect of elevated CO_2 on marine organisms have been critical for assessing the potential impacts of ocean acidification. Likewise, the Free Air CO_2 Enrichment (FACE) studies have helped us to better understand CO_2 fertilization in the terrestrial environment, including the constraints on increased net primary production and the potential for carbon storage in soils.

Despite such progress, different responses observed across studies can be hard to attribute to changes in CO_2 as opposed to interactions with other environmental variables or to differences in experimental protocols. A reconciliation of the observed responses to manipulative experiments is particularly important for distinguishing natural variability from human-induced changes in the terrestrial and aquatic carbon cycles.

In addition to manipulations of the physical and biological environment, manipulative experiments may be useful to test human decision making associated with the carbon cycle and carbon incentives. Although manipulation of human systems needs thoughtful consideration, different states and countries are pursuing multiple policy paths that can be evaluated. As is the case for differences in the physical environment, differences in economic, cultural, and social environments and access to information offer opportunities to test these manipulations. Such manipulative studies could include research on human behavior, natural resource economics, and other areas of socioeconomic systems related to the carbon cycle. *Thus, manipulation experiments with common protocols that span broad environmental and socioeconomic gradients, and that simultaneously manipulate multiple factors, are needed. The focus of these studies should be to understand the responses of the net carbon balance to changing environmental conditions and socioeconomic drivers.*

Some cross-disciplinary process studies and multi-factor manipulative experiments should be located in regions where the responses to manipulation are likely to reveal vulnerabilities in carbon storage (e.g., permafrost ecosystems) or are used to assess large-scale carbon management strategies, such as carbon capture and storage (CCS) underground or under the ocean floor. Geological sequestration programs combine interactions between social policy and scientific research for climate change mitigation; thus, interagency resources should be prioritized to develop integration among these efforts as much as possible. In order to identify system vulnerabilities, these studies must not only focus on systems that have already been identified as having potential tipping points in their ability to store carbon, but should also include systems that have yet to be examined. Given finite resources, it will be important for the scientific community to prioritize the focus of a new generation of process studies and manipulative experiments.

5.2.3 *Integrative field campaigns*

Integrative field campaigns provide intensive data, and have the ability to test different approaches for examining carbon cycling at a range of nested scales. The Interim synthesis activities currently being conducted through the North American Carbon Program are providing a first, key opportunity to coordinate field campaigns and modeling efforts to reconcile understanding of carbon cycling for particular systems across a range of nested scales. The most mature of these is currently the Mid-Continent Intensive (NACP, 2010a), which represents a coordinated effort of field and airborne observations, atmospheric observations, inventory development, biospheric modeling, and atmospheric inverse modeling to improve understanding of carbon cycling in the agricultural Midwest region.

Related efforts focusing on coastal and other systems are currently beginning. These types of efforts will provide a key opportunity to synthesize understanding gleaned through carbon cycle studies over the past several decades by providing a platform for integrating different types of data across different spatial and temporal scales, and obtained through a variety of mechanisms.

Integrated field campaigns are needed more generally, supported independently of focused observational studies and modeling, to provide a clear opportunity for synthesis across the carbon cycle science community. One specific opportunity is to have integrated field campaigns include tests of human decision making and other socioeconomic factors of how people affect the carbon cycle. For instance, the Large-scale Biosphere-Atmosphere Experiment in Amazonia examined how Amazonia functions as a regional entity within the larger Earth system as well as how changes in land use and climate affect the biological, physical, and chemical functioning of the region's ecosystem. Socioeconomic factors have not traditionally been part of similar research campaigns. In the future, campaigns should combine critical data needs in the physical and biological sciences with relevant socioeconomic data in the same locations. For instance, campaigns could quantify the human factors that drive changes in land use, fish catches, or other factors relevant to productivity and the carbon cycle. More integrative field campaigns are needed to test different approaches for examining carbon cycling at a range of nested scales. Such intensive studies will also be useful for elucidating the dynamics of anthropogenic emissions, such as through urban-scale field campaigns.

5.2.4 Mapping studies of system dynamics and functions into the goals

The studies in this element contribute significantly to all of the stated goals. Understanding processes is a key part of providing a clear and timely explanation of the mechanisms behind past and current variations in atmospheric CO_2 and CH_4, as stated in Goal 1. These studies will also help determine realistic uncertainty estimates by placing bounds on mechanisms and responses. An understanding of system dynamics is central to understanding the socioeconomic drivers of carbon emissions in Goal 2. New socioeconomic components of Element 2 include consideration of manipulative experiments to test human decision making for carbon incentives and extending traditional gradient studies from biophysical variables to socioeconomic ones. Goal 3 calls for an understanding of vulnerabilities and prediction of future carbon cycle changes that can only be achieved with a process-level understanding of biological, physical, and socioeconomic systems. Process studies are also central to interactions of the changing carbon cycle and climate with ecosystems, biodiversity, and natural resources, the core of Goal 4. The carbon cycle is intimately linked to ecosystem services that people value, including water resources and biodiversity. The theme of ecosystem services extends to Goal 5, which examines the side effects of carbon management pathways, another area for which process-level observations and manipulations are central. Process measurements associated with carbon management experiments will provide mechanisms and likely outcomes for different carbon management alternatives. Finally, Goals 5 and 6 together, including the needs of decision makers for carbon cycle information, require an improved understanding of human choices and the responses of the natural environment to those choices. These can only be achieved with an understanding of integrated system dynamics.

5.3 Modeling, prediction, and synthesis

Numerical and statistical modeling have been crucial components of carbon cycle research over the past decade and will continue to play a central role over the next decade. Modeling studies provide unique opportunities for data analysis, mechanistic exploration, and prediction of human and natural interactions across spatial and temporal scales. Models integrate theory and observations to enhance our understanding of the carbon system, and they will also form the backbone for synthesis. Inter-comparison activities that merge inventory and site-level data with the mechanistic relationships embodied in the models themselves provide the opportunity to understand core processes and to evaluate the uncertainty in our understanding. These syntheses push forward our process-level understanding of the carbon cycle and are critical for identifying knowledge gaps.

In each of the elements described in this section, particular emphasis needs to be placed on quantifying and, where required, reducing uncertainty. In the current context, quantifying uncertainty goes beyond assigning traditional error bars that represent the sensitivity of model results to specific parameters or model-generated uncertainties based on statistical assumptions built into the modeling framework. *Instead, significant effort needs to be invested in developing methods for uncertainty quantification that reflect all sources of uncertainty affecting a particular model estimate.* Such objective levels of scientific understanding can be used to communicate scientific results quantitatively, or even qualitatively, both within and beyond the carbon cycle science community. Once appropriate tools are available for uncertainty quantification, the process of uncertainty reduction through model development, improved observations, and model evaluation becomes a meaningful way of tracking the evolution of the state of the science. Characterization of uncertainty is also important to informing policy and decision making.

In the realm of global change, there is great interest in predicting and anticipating the future. Modeling and synthesis activities are critical to our skill in predicting the future, but there is always inherent uncertainty in any predictions. There is a need to develop and present realistic and useful ways to convey the uncertainty of projections, the range and distribution of potential outcomes, and the consequences of uncertainty.

5.3.1 Improve existing models

In the next decade, model skill needs to be improved through enhanced collaboration among field, laboratory, and modeling scientists. Additional effort will need to be invested in developing models that bridge and synthesize information across traditional disciplines, scales, and data types. Particular emphasis needs to be placed on analyses and theoretical developments that provide the necessary bridge between field and remote-sensing observations and fundamental understanding of the carbon system. Process studies (Element 2) will provide critical mechanistic information that can be used to improve model parameterizations. In addition to improvements in mechanistic understanding of individual components of the carbon cycle, improved models are needed to link between components, such as coupled land-(coastal) ocean models at regional scales to better understand areas where land, ocean, atmosphere, and humans interact at very small scales. In addition, targeted programs promoting model-model and model-data comparison, including benchmarking of regional and global models, can be particularly useful to identifying fundamental knowledge gaps.

Unlike the carbon cycle models that focus on one or more components of the carbon system described in the last paragraph, Earth System Models attempt to fully represent the evolving climate system and to predict its future state. Such models are used, for example, in the IPCC efforts. These models can be used in hind-cast mode, looking backward, to represent the past climate, and then their predictive success can be tested through comparisons to historic data. The complexity of Earth System Models makes them enormously computationally intensive. There are some model processes that can best be improved by increasing model resolution and there are additional, highly complex modules that need to be included. Thus, we should expect Earth System Models to continue to stress computational resources in the next decade. Innovative computing strategies should be pursued that will make this software run faster or make the hardware more powerful.

Furthermore, as we move into the era of Earth System Modeling in which carbon and climate modules interact, allowing for feedbacks between climate and carbon cycling, the interpretation and attribution of uncertainty bounds and confidence limits will become increasingly complex and challenging. Focused research on uncertainty estimation involving scientists from multiple communities will be needed to help interpret Earth System Model-based analyses and predictions.

In order to explain observed variability and trends in the atmospheric CO_2 concentration (Goal 1) and track carbon emissions and sequestration activities in land and in aquatic systems (Goal 2) we must have optimal diagnoses of the current state of the carbon cycle and the mechanisms driving variability and change. *This calls for the development of models that allow for the integration of multiple sources of data, and to integrate information across spatial and temporal scales.* The continued evolution of parameterization approaches for process-based models, the ongoing improvements to the statistical framework of inverse models, and the ability of all models to ingest a wider variety of *in situ* and remote-sensing data types (as well as the recent early steps in integrating computational approaches based on numerical data assimilation) are all providing opportunities for such explicit integration of data across types and scales. Much work will be needed over the next decade to develop the necessary conceptual, numerical, and statistical tools to fully benefit our community and to make optimal use of the expanding set of observations available for informing our understanding of the carbon cycle.

5.3.2 *Add human dimensions to Earth System Models*

Integrated Earth System Models are extremely complex because they ideally strive to include the entire range of processes and feedbacks across the spectrum of human and natural processes. Development of these models in the last decade has focused on physical parameterizations and great strides in representation of biogeochemistry have also been made. These models must continue to be developed, as outlined above. *In addition, however, more complete and complex representations of human activity are needed in Earth System Models.* Trends and distribution patterns in demographics, migration, international trade, economic development, human settlements, world view, transportation technology, agricultural practice, and materials substitution only begin to enumerate (much less quantify) the multitude of factors that will impact and/or be impacted by the global carbon cycle and efforts to manage the human perturbation of the carbon cycle. The available alternatives, advertent and inadvertent incentives, and choices faced by people and the feedbacks from the climate system will have huge impacts on the path of change in the global carbon cycle. It is increasingly important to know what we can know and to deal with what we do not or cannot know. Modeling and analytical systems can begin to represent the interactions and side effects of the multitude of interacting factors and to identify the critical parameters and inter-linkages.

Finally, the needs of decision makers need to be considered as models are developed and simulations are planned. The scientific community has the opportunity to contribute to the need for carbon/climate management through a more coordinated Earth System Model development effort. Interactions with decision makers are needed both to frame the questions for research and to pursue the answers from research. These efforts will require new frameworks and centers for trans-disciplinary collaboration as well as a renewed commitment to data management and computational resources.

5.3.3   Augment synthesis activities

Synthesis brings together data products and models that attempt to capture similar or related processes and evaluates the degree of agreement between the different data and modeling approaches. Synthesis efforts are critically important to identifying gaps in knowledge and for leading to new studies to fill those gaps. However, synthesis activities are often difficult to fund using classic mechanisms, largely because they are generally not 'new' science. Funding strategies to support critical synthesis activities are needed.

The 1999 U.S. Carbon Cycle Science Plan led to the North American Carbon Program. The NACP's current Interim Synthesis is a good example of the power of the synthesis and inter-comparison process. Currently, as part of the NACP's interim site synthesis activities (NACP, 2010b), tens of models are being compared to data at tens of locations. In addition to enhancing knowledge of terrestrial carbon cycle processes, this activity is making it possible to identify critical differences among models. This process could not occur without facilitation by the Modeling and Synthesis Thematic Data Center (MAST-DC) (NACP, 2010b), a dedicated carbon cycle modeling and data synthesis center that is standardizing formats and providing repositories for community data sets and model results. Similar to NACP's current Interim Synthesis effort, coordinated synthesis efforts are needed for the open ocean, the coastal zone, the entire land to deep ocean system, and integrated human-natural assessments. For example, early results from the NACP Site Level Data-Model comparison project (Schwalm et al., 2010) illustrate how far we are from having yet developed a terrestrial carbon cycle predictive capacity.

5.3.4   Mapping of modeling, prediction, and synthesis into the goals

Modeling and synthesis are essential for the successful achievement of all six goals of the Plan. Many kinds of models will contribute to the goals, including process-based models, atmospheric and oceanic inverse models, flux inventories, and tools for numerical data assimilation that are increasingly being used to parameterize models. Models and simulations already help to explain variability and trends in past and current atmospheric CO_2 and CH_4 concentrations (Goal 1) and project greenhouse gas concentrations into the future. Models are also needed to track the socioeconomic drivers behind carbon emissions and sequestration (Goal 2). Vulnerabilities in carbon stocks, such as permafrost loss and critical feedbacks with biodiversity and other ecosystem services, all need to be characterized through experiments and model simulations (Goals 3 and 4). In addition, interactions across the scientific and decision-making communities will be critical for analyzing the efficacy of carbon management strategies and for the effective use of analyses and predictions (Goals 5 and 6). Although modeling of the physical environment, in particular, has made tremendous strides over the last decade, models that more accurately represent biologically driven processes and human interactions are needed to address complex carbon cycle interactions and to quantify accurately the current uncertainties associated with our understanding of components of the global carbon system.

5.4   Communication and dissemination

Effective communication and dissemination of the results of carbon cycle science research are essential if the investments made in science are to become useful in other studies and in informing decision making and conservation efforts. Communication is, of course, a two-way street and, in this time of increasing concern about climate change, it is important that scientists not only communicate their results but that they are receptive and attentive to the output of other disciplines and the needs of decision makers and the general public. Traditional means of communication such as publication in the peer-reviewed literature is a necessary but not sufficient vehicle for communicating the results of research. Research suggests that boundary organizations or individual 'integrators' can be effective in bridging between scientific organizations and decision makers, but this function needs to be deliberately created and fostered (Buizer et al., 2010). Communication and dissemination of science is a confusing issue at times because it is not necessary or appropriate that every scientist reach out and communicate with the public, but it is necessary that some scientists bridge the gap between the laboratory and peer-reviewed technical literature and the decision makers and public who use the results and support the conduct of science.

An active program of communication will enhance the impact and responsiveness of the research conducted within the scope of the Plan. *These aspects include an emphasis on two-way communication with the broader community, the translation of scientific results into information that is directly usable by related communities (in keeping with Goal 6), and the promotion of better understanding by the scientific community of the decision-making process (also in Goal 6).* A successful communication strategy will include assessing uncertainties about information and conveying those in a way that is relevant and understandable to users (e.g., Morss et al., 2008).

5.4.1   *Improve dialogue among the decision-making community, general public, and scientific community*

For many, the term communication conjures up images of pamphlets, websites, or lectures designed to bring the most important results of research to a wider audience. Communication research shows, however, that this sort of passive, one-way communication is rarely successful at promoting the integration of new knowledge into practice so that it can make a difference and change outcomes (e.g., Lemos and Morehouse, 2005; Moser and Dilling, 2007). Communication and dissemination strategies can fail on many grounds – because the information is not relevant to a decision, because the information presented is confusing or not understandable to the audience, or simply because the channel used to disseminate information is not well attended by the intended audience. When addressing an international audience, cultural perspective can also complicate effective communication of information. These findings suggest that it is worth the time up front to think carefully about how to craft a communication strategy and to invest in research to refine communication efforts such that they can be most effective.

Perhaps one of the most common truisms in communication is the exhortation 'know thy audience.' Only by knowing the intended audience will we know what matters to them. Carbon cycle science is a relatively complex subject to discuss,

and yet one that intersects every person's life through the decisions he or she makes. The research strategy outlined in Goal 6 is designed to develop an understanding of what information is relevant to particular groups of decision makers. The challenge of a communication strategy is to ensure that the information is then appropriately disseminated and understood in order to fully support decision making. This element will work with all goals but especially Goal 6 in order to ensure an effective outcome.

As part of this element, the scientific community needs to work in tandem with communities outside the carbon cycle science research community both in the United States and abroad to identify the target audience for and potential users of the results of carbon cycle research. There may be several audiences, from sophisticated policy analysts to individuals trying to understand voluntary offset programs. In addition, *the implementation of this Plan needs to create a communication focus up front as an ongoing part of the program, and to ensure that this focus is maintained throughout the implementation of the Plan.*

5.4.2   *Develop appropriate tools for communicating scientific knowledge to decision makers*

It is imperative that the research conducted as a result of this Plan yield tools that translate results from scientific synthesis and prediction into quantitative, understandable products for policy and management professionals. Knowing the ultimate audience for scientific results means knowing how they access and understand new information. All too often scientific programs tend to have a 'loading dock' mentality about the production of information, which entails creating the information, putting it on the loading dock, and assuming or hoping that someone will come and pick it up (Cash et al., 2006). The 'loading dock' of science is usually peer-reviewed journals, which are necessary for the process of science, but which are also largely inaccessible to the majority of decision makers and even less so to the broader public or international groups. Other mechanisms of reaching decision makers with information must be implemented, including different types of trade meetings, workshops, newsletters, and outreach networks. Program managers should also work with stakeholders to design carbon cycle requests for proposals that specifically address manager and decision maker needs for information.

5.4.3 Evaluate impact of scientific uncertainty on decision makers

New approaches are needed for effectively communicating the level of certainty that scientists have in various components of the carbon budget to managers, decision makers, and the general public. The graphics and flowcharts that are currently typical of carbon cycle science are not intuitively understandable to the general public. The carbon cycle science community has an opportunity to more effectively communicate scientific uncertainty to decision makers by engaging communications researchers from other scientific communities. These communities include the weather and climate research and forecast communities. For example, researchers in the climate arena have studied for years how to present the notion of probabilities in such a way that forecasts can be properly understood in the context of the range of uncertainty, rather than as a deterministic single prediction of the future. The graphics used in endeavors such as drought monitoring, weather forecasting, and climate forecasting are tested and refined based on experience and feedback from users As carbon cycle science becomes increasingly relevant to decision makers, this community will need to develop and test creative ways of reporting findings so that the intended meaning is received.

As effective methods for communicating uncertainty are developed, the carbon cycle science community must also work together with the broader public to evaluate the implications of uncertainty in present-day knowledge and in future carbon cycle projections.

5.4.4 Mapping communication and dissemination into the goals

The communication and dissemination activities described in this element contribute to all of the stated goals. The clear and timely explanation of past and current variations observed in atmospheric CO_2 and CH_4 (Goal 1) includes effective communication with communities beyond carbon cycle science as well as the development of appropriate tools for communicating scientific knowledge and its associated uncertainties. Understanding and quantifying the socioeconomic drivers of carbon emissions and developing transparent methods to monitor and verify those emissions (Goal 2) also requires the establishment of a two-way dialogue between the carbon cycle science community and

decision makers, as well as the development of tools for effectively communicating scientific understanding with this community and learning the questions being confronted by this community. The ability to learn and evaluate the impact of uncertainty on decision makers can inform research on determining and evaluating the vulnerability of carbon stocks and flows to future climate change and human activity (Goal 3). Similarly, the goal of predicting how ecosystems, biodiversity, and natural resources will change under different CO_2 and climate change scenarios (Goal 4) will have a larger impact if appropriate tools for communicating scientific knowledge are available. If we are to determine the likelihood of success and the potential for side effects of carbon management pathways (Goal 5), it is essential to understand and evaluate the impact of scientific uncertainty on this decision-making process. Finally, the goal of addressing decision maker needs for current and future carbon cycle information, and providing data and projections that are relevant, credible, and legitimate for their decisions (Goal 6), is directly linked to the ultimate goal of effective communication and will support more targeted, effective communication efforts.

5.5 Resource requirements

The current state of carbon cycle science, together with the four research elements identified above, begins to outline a coherent, integrated research program with priorities that include both existing and new components. Significant new resources are needed to continue with high priority, existing initiatives; to reach out in important new directions; and to accelerate progress toward confronting a problem whose relevance to human welfare is becoming increasingly apparent. Our ability to reach the stated goals within the next decade will depend heavily on the ability of the U.S. funding agencies to provide full financial support to the Plan. The 2009 edition of *Our Changing Planet* report to Congress estimated that the total U.S. carbon cycle science budget, excluding platform costs (e.g., satellites, ship time, aircraft time), is currently about $170 million per year (US Global Change Research Program, 2009a). While this budget has grown substantially over the last five years, many of the observing networks, process studies, and model development efforts are at less than half of the levels necessary to understand the global carbon cycle at the required level of detail for effective management decisions. Meaningful integration of the human dimensions components of the carbon cycle has not been possible with the current level of funding, and very few resources have been devoted to communication and dissemination, making the path and time required for new research findings to inform management decisions extremely inefficient.

Many of the planning documents listed in Chapter 2 include detailed budgets for the implementation of their components of carbon cycle research. Using these detailed budgets, the budget compiled for the 1999 Science Plan, and an assessment of what the carbon cycle community has accomplished with the current levels of funding, we estimate that the total U.S. carbon cycle budget, excluding platform costs as listed above, will need to be increased to approximately three times the current investment of $170 million, or roughly double the $250 million estimates presented in the 1999 Science Plan. This yields an estimate of $500 million per year to achieve the goals outlined in this Plan. Note that, as with other budget estimates, including the USGCRP carbon cycle science budget and the 1999 Science Plan, this estimate does not include the full cost of leveraged assets, such as the platform costs outlined above, as well as budgets of existing program in related fields that can be leveraged to working at the interface of the carbon cycle community and other scientific groups. A detailed breakdown of the costs for specific components will have to be determined through the process of developing implementation plans that are based on this Plan. At this phase, however, we do propose an approximate distribution of the proposed carbon cycle funds into the four elements described in this chapter:

- $175 million for Element 1, Sustained Observations, which is similar to the proposed 1999 Science Plan funding level
- $125 million for Element 2, Process Studies, which is approximately three times the 1999 Science Plan funding level, to include ecosystems and human dimension studies

- $150 million for Element 3: Modeling, Prediction, and Synthesis, which is substantially higher than the 1999 Science Plan funding level because of previously underestimated costs for synthesis and data management, as well as support for many more modeling groups needed to address decision maker needs
- $50 million for Element 4: Communications and Dissemination, which was not included explicitly in the 1999 Science Plan, but needs a significant infusion of startup funds

Chapter 6

Interdisciplinary and International Collaboration and Cooperation

This Plan recognizes the critical role that carbon cycle science is currently playing as interest in and concern about climate change become central issues in human considerations. Carbon cycle science plays a central role in understanding the Earth's climate system and in human considerations of how to manage long-term global change. Carbon cycle science is not an arcane subject at the margin of human affairs, but is rather a topic of general interest and of great need for environmental, economic, and human health. The boundaries of carbon cycle science are no longer just biogeochemical, but include linkages to the environmental, economic, and social sciences and to international affairs and public policy. The system boundaries of what is considered to be 'carbon cycle science' are blurring as we recognize these linkages and the need for collaboration, cooperation, and the sharing of ideas and research results. Research in the traditional atmospheric, marine, and terrestrial aspects of the carbon cycle remain vital, but we are increasingly aware that there are human dimensions to the carbon cycle and that carbon cycle science impinges directly on issues of economics, engineering, and public policy. As managing the carbon cycle becomes increasingly common, institutional issues become important and the need for understanding and dealing with uncertainty becomes central.

The geographic boundaries within carbon cycle science are similarly blurring. What happens in the boreal tundra, the tropical forests, the coral reefs, and the global ocean affects us all – and these ecosystems are in turn affected by all of us. It is increasingly apparent, also, that issues of the global carbon cycle are global in every sense of the word and that research supported by U.S. interests needs to be cognizant of and collaborative with research efforts around the world. Examples of key international carbon cycle organizations include the Global Carbon Project and the Group on Earth Observations, coordinating efforts to build a Global Earth Observation System of Systems.

The United States has been, and must continue to be, a leader of and major contributor to international research efforts; but the United States cannot bear the full burden and needs the expertise of other scientists around the world. Measurement and monitoring of global change need broad global effort and the full range of carbon cycle studies finds able colleagues and collaborators in many countries. The variety of social, cultural, and legal systems on Earth create a diversity of interests for understanding the human impact on climate and the human capacity for mitigating or adapting to changes in the carbon cycle. The U.S. educational system plays a major role in preparing the world's science community.

6.1 Interdisciplinary collaboration and cooperation

The challenge today is to broaden and redefine the boundaries of carbon cycle science; to build the linkages with research areas with common, interdependent, or overlapping concerns; and to ensure that critical topics or data needs do not drop into cracks between traditionally defined disciplines or sources of research support. We need to ensure that there is support for interdisciplinary and cross-disciplinary studies, and that human elements are incorporated into carbon cycle studies whenever appropriate. Interdisciplinary studies and improved linkages with the social and political sciences are essential, and visions of the future need to be strengthened through interactions with integrated assessment studies and studies of carbon management. Research related to the development of biomass fuels, Reducing Emissions from Deforestation and forest Degradation (REDD), and the impacts of urbanization on coastal processes and marine resources are clear examples in which the human dimensions need to be a critical part of basic enquiries.

The future will require improved interaction and information exchange not only within and among different scientific disciplines, but also with stakeholders and decision makers – people who require up-to-date assessments, improved approaches for understanding complex and interdependent issues, and ways of quantifying and dealing with uncertainty. There is a need to bridge the differences between the research results published by scientists and the information needed by stakeholders and decision makers – to translate research findings into meaningful input for these groups. An ongoing dialogue among the different groups is needed to raise awareness of both what science can provide and what science cannot provide, and of the uncertainties associated with current assessments and projections of the future.

6.2   International collaboration and cooperation

The increasing importance of international collaboration is also apparent. This Plan is envisioned for the U.S.-based research community but the carbon cycle and its impacts are global and important studies are needed, for examples, in tropical and boreal areas, in the global oceans, and in other cultures. Observational networks, mitigation and adaptation strategies, predictions of the future, verification of commitments, etc., will all depend on participation and contributions from scientists around the world. The United States contributes importantly to international research efforts through its educational system and through access to unique information such as that provided by remote-sensing satellites and research vessels. U.S. scientists need to participate and take leadership roles in international assessments and syntheses, field campaigns, model inter-comparisons, and observational networks. All of this international participation offers opportunities to capitalize on other resources and to contribute the knowledge and creativity of U.S. scientists to coordinated research. Because of the benefits of international collaboration, this coordination should be realized for the full

cycle from program planning to project execution and data management.

While many international programs will involve large projects and efforts and multiple participants, opportunities for productive efforts also exist at the scientist-to-scientist level where cooperation with a single institution or exchanges of staff or students can facilitate or accelerate progress in critical geographic or disciplinary areas. U.S. scientists and students need to study abroad and U.S. institutions need to provide opportunities for scientists and students here.

There are numerous projects, opportunities, and initiatives wherein researchers from many countries share data, insights, manpower, and platforms such as ships and satellites. We hesitate, for fear of important omissions, to list the many successful international research efforts currently focused on cooperation and collaboration among scientists from multiple countries, but we encourage thoughtful evaluation and U.S. participation whenever useful and when the whole is greater than the sum of the parts. U.S. scientists should be encouraged to take on leadership roles in these efforts. The IPCC can be cited as an example of scientists from around the world sharing information and views with the result of broad synthesis and important input to the decision-making process. There are many additional, more focused, collaborations that provide essential resources, synthesis, or intellectual support and where win-win situations result in the United States both giving and receiving important understanding.

Another important aspect of international cooperation involves sharing of data sets required for carbon cycle science. Current efforts to increase free public access to satellite observations, for example, are proceeding very slowly, and in some instances, access has been reduced over a period of several years.

Redoubled efforts on the part of the U.S. government and the international community are needed to reform data policies. This issue is likely to become an increasingly acute barrier to scientific progress as new international satellite missions provide important new constraints on carbon cycle processes, including ecosystem structure and land cover/land use change. Transparent access to all of the data used for carbon cycle monitoring internationally is needed to ensure the success of climate mitigation efforts.

6.3 Supporting and stimulating collaboration and cooperation

Research that is sensitive to the role of humans in the global carbon cycle and the needs of people for carbon cycle science can be facilitated by interagency working groups and advisory panels that include strong participation by human dimensions researchers and social scientists. Specific research calls or pilot projects can be designed to encourage or require interdisciplinary collaboration beyond traditional alliances and interactions. Growing research initiatives should make a pointed effort to ensure that human components and human needs are incorporated from the beginning.

In order to encourage or facilitate the kinds of inter-disciplinary, international cooperative and collaborative projects needed, it may be appropriate for U.S. funding agencies to try innovative support structures, support targeted workshops, or issue specific calls for proposals that focus on areas in need of creative ideas to make progress. The kinds of mingling of ideas that are needed to solve interdisciplinary and global-scale problems are often non-traditional and specific measures may be needed to create the required dynamics.

Chapter 7
Implementation and Funding

7.1 Integration of program priorities

Throughout this document we have articulated long-term priorities for a new U.S. Carbon Cycle Science Plan, including the overriding scientific questions that drive this new Plan (see Chapter 3). The research goals in Chapter 4 represent current research priorities that are derived from these long-term priorities, and that should be achievable within a 10 to 20 year timeframe, assuming that sufficient funding is available (see Section 7.3). The elements described in Chapter 5 cut across all of the goals and define priorities for specific, actionable components of the research plan. Each of the four top-level elements contributes to all six research goals, thus defining the cross-cutting building blocks needed to address the fundamental science questions outlined in this Plan. Carbon cycle research is central to understanding fundamental Earth systems and developing a sustainable future, and the overriding research priority is for a balanced, integrated program that addresses the multitude of interconnected and critical questions that remain.

Building on Chapter 5, which addressed how the individual elements build towards the goals outlined in Chapter 4, we present a complementary view in Table 7.1, examining how each goal draws on these elements. As the table indicates, each of the six goals relies on at least some components of each of the elements. While the text within the table in no way represents the full extent of the connection between the goals and the elements, this text clearly illustrates the interconnectivity of all the components of the Plan. Selectively funding specific components of the Plan while underfunding other components would degrade progress on all of the

goals. Ultimately, detailed implementation plans will need to be developed for each of the six research goals but they are beyond the scope of this document. These implementation plans will need to consider the proper balance of elements based on the total level of support available.

7.2 Implementation opportunities and barriers to success

This report describes new priorities for carbon cycle science that complement sustained, ongoing priorities and research. Throughout this Plan, we have also provided suggestions for how to implement these new priorities and integrate them with current research and funding mechanisms. In addition, however, the conduct of science depends on the institutions and institutional structures that support the research, and this brief section provides additional recommendations for institutional structures to improve coordination and to ensure the achievement of the Plan's research recommendations.

- *Provide more opportunities for sustained, long-term funding.* Many aspects of carbon cycle science cannot be adequately maintained on a three- to five-year funding cycle. As one example, long-term observational efforts for the atmosphere, land, and oceans need funding continuity and certainty. In addition, where new long-term measurement and observational efforts are planned or underway, such as through the NSF National Ecological Observatory Network (NEON) or the Ocean Observatories Initiative (OOI), carbon cycle scientists should be central to such efforts and able to participate in a sustained fashion.

- *Enhance Carbon Cycle Data Management.* The creation of an integrated carbon data management system and distributed repository would make carbon data more accessible and directly usable both across and within disciplines. Such a system should include not only observational data, but model drivers and outputs, and, to the extent possible, open-source versions of the models themselves. This system should also include a consistent data use policy across federal agencies to facilitate data interchange. U.S. data should be freely and easily accessible and every effort should be pursued that the same is true for data from other research programs.

- *Encourage directed calls for integrated topics in carbon cycle research.* In addition to regular, disciplinary funding; additional targeted calls to link research in the social, physical, geochemical, and biological sciences could provide a framework for advancing integrative carbon cycle science. The integrated calls should cover all aspects of research described in this plan that require substantial cross-disciplinary collaboration. As an example, calls could be included to explore specific carbon cycle vulnerabilities in physical systems such as tundra permafrost or in socioeconomic systems that drive carbon cycle losses (including losses through deforestation). Similarly, a funding call could

Table 7.1a: Mapping of the six science goals onto the first two of the four program elements.

	E1: Sustained Observations			E2: Studies of System Dynamics and Function Across Scales		
	Atmospheric; ocean, coastal and inland-water; and terrestrial observations	Monitoring and assessment of human systems	Remote-sensing observations of Earth system	Intensive process studies and field campaigns	Manipulative lab and field studies	Integrative field campaigns
G1: explain variations in atmospheric CO_2 and CH_4	document variations	helps explain drivers for variations	document variations	can confirm explanations of variations	investigate processes to help explain variations	provide link between process studies and observations
G2: Understand drivers and quantify emissions	document impact on atmosphere, aquatic systems, and carbon stock changes	informs socio-economic drivers of anthropogenic emissions	document impact on atmosphere and some stock changes	test theories on human drivers and quantification methods for emissions	test theories on human drivers and quantification methods for emissions	test approaches for quantifying emissions
G3: Evaluate carbon vulnerability	document changes in stocks and flows and impact on aquatic systems	determines which stocks and flows may be impacted by humans	document changes in stocks and flows	assess vulnerability	test specific vulnerabilities	assess vulnerabilities at regional scales
G4: Predict ecosystem changes	document ecosystem, biodiversity, and natural resource changes	informs possible carbon/climate scenarios	document ecosystem, biodiversity, and natural resource changes	assess ecosystem responses and feedbacks	test ecosystem responses and feedbacks	assess responses and feedbacks at regional scales
G5: Evaluate carbon pathways	document impacts of current pathways	informs likely management pathways	document impacts of current pathways	assess impact of current pathways	test specific results of management decisions	assess impacts of pathways at regional scales
G6: Address needs for information	provide global information for decisions, and document stock and ecosystem changes	provides information on decision maker needs	provide global information, and document stock and ecosystem changes	provide information on local carbon cycle processes	provide information on impact of specific processes	provide information on regional carbon cycle processes

Table 7.1b: Mapping of the six science goals onto the second two of the four program elements.

	E3: Modeling, Prediction, Synthesis			E4: Communication, Dissemination		
	Improve existing models	Add human dimensions to Earth System Models	Augment synthesis activities	Establish dialogue among communities	Tools for communicating scientific knowledge	Evaluate impact of uncertainty on decision making
G1: explain variations in atmospheric CO_2 and CH_4	quantifies fluxes and represents processes controlling variations	links human and natural processes	reconciles understanding across observations/models	integrates understanding across communities	helps reach a broader community	identifies uncertainties that must be addressed
G2: Understand drivers and quantify emissions	improves emission assessments and prediction of socio-economic drivers	links humans with natural processes	reconciles understanding across observations/models	integrates understanding across communities	helps reach a broader community	identifies uncertainties that must be addressed
G3: Evaluate carbon vulnerability	evaluates future responses	links humans with natural processes	reconciles understanding across observations/models	integrates understanding across communities	helps reach a broader community	identifies uncertainties that must be addressed
G4: Predict ecosystem changes	evaluates future responses	links humans with natural processes	reconciles understanding across observations/models	integrates understanding across communities	helps reach a broader community	identifies uncertainties that must be addressed
G5: Evaluate carbon pathways	evaluates future responses	links humans with natural processes	reconciles understanding across observations/models	integrates understanding across communities	helps reach a broader community	identifies uncertainties that must be addressed
G6: Address needs for information	gives process diagnosis, attribution, and prediction	links humans with natural processes	reconciles understanding across observations/models	integrates understanding across communities	helps reach a broader community	identifies uncertainties that must be addressed

extend traditional research along biogeochemical environmental gradients to include gradients in socioeconomic factors.

- *Facilitate efforts to contribute to integrated, interdisciplinary efforts such as the assessments of the Intergovernmental Panel on Climate Change.* In the past decade, interactions between the carbon cycle and the physical climate system have emerged as a primary source of uncertainty in projections of global change into the 21st century. The development, evaluation, and improvement of the carbon cycle scenarios used in efforts such as the IPCC assessments should be supported as a community-wide effort, and should reflect the best and most current science from the carbon cycle research community.

- *Establish stronger links between the Carbon Cycle Interagency Working Group (CCIWG) of the U.S. Global Change Research Program and other U.S. interagency working groups focused on climate change and mitigation.*

Of the ten IWGs, the CCIWG has been particularly active over the past decade. Most of the other nine, including Atmospheric Composition, Ecosystems, Land Use and Land Cover Change, and Observations and Monitoring, have elements with common or interacting components to carbon cycle research. As a result, increased cooperation among these groups could enhance the success of the new carbon-cycle plan. Similarly, closer ties between the Environmental Protection Agency's greenhouse gas inventory team and the CCIWG may accelerate progress in reducing uncertainties from key sectors of the U.S. carbon budget.

- *Develop a strong connection between carbon cycle research and the developing ocean acidification program.* Recent legislation has led to substantial new investments in ocean acidification research and the potential development of a national ocean acidification research program. As the ocean acidification program develops, it will be important to link the ocean acidification efforts with more generalized carbon cycle research to leverage these efforts and take advantage of potential synergies.

- *Expand the North American Carbon Program to a new Northern Hemisphere Carbon Program.* The North American Carbon Program was established a decade ago to measure and understand the sources and sinks of carbon dioxide, methane, and carbon monoxide in North America and in adjacent ocean regions. To resolve uncertainties in the global carbon sink this decade, a new emphasis is needed on the Northern Hemisphere *in total.* The expanded emphasis would allow scientists to reconcile observations with model estimates of carbon cycling in the Northern Hemisphere.

- *Improve international linkages.* Carbon cycle research requires a global focus that in many cases is best served through collaborative studies with international partners. Although scientists are generally interested in working together with their international colleagues, funding and legal restrictions often make these interactions difficult. The U.S. agencies should explore ways of promoting and facilitating international collaborations, including reforming data policies to increase access to satellite observations from other countries.

- *Use the North American Carbon Program and Ocean Carbon and Biogeochemistry program as models to initiate similar, problem-oriented research communities.* The reach of global carbon cycle research is sufficiently interdependent and broad that it is important to achieve interaction, and yet this is difficult to do within specific projects or disciplines. The NACP and OCB have succeeded in establishing on-going dialogue, interaction, and shared learning that have benefited its many diverse constituents. One group with strong roots in both the social and natural sciences is suggested. With guidance from the carbon cycle science community and the research priorities articulated in this Plan, funding agencies should seek to identify opportunities for other large-scale efforts, using NACP and OCB as successful models.

- *Implement a process for periodic measurement and evaluation of progress in pursuing the goals of this Plan.* This Plan should not be perceived as a one-time, static statement. There should be a commitment to periodic (perhaps every three years) examination to evaluate the extent to which the goals are being achieved and to ensure that the goals outlined here remain appropriate in a rapidly evolving scientific, environmental, economic, and political environment.

- *Continue to provide broad support for education and training.* The U.S. research agencies have long provided support for education and research involvement by students and early career scientists at all levels and by under-represented groups. We endorse this as a wise investment in the future of our science and strongly encourage continuation. In addition, there is a need for an increased emphasis on interdisciplinary education focusing on carbon/climate science and decision making in a global context. A future low-carbon economy will require a new workforce of engineers, scientists, economists, lawyers, policymakers, and financial experts who understand that climate-related decisions have global impacts. New educational efforts will be needed to build the necessary skills and systems thinking approaches to carry out the energy transformation needed to overcome the climate challenge.

7.3 Program support

The principal priority detailed in this research Plan is to develop and maintain a broadly focused, balanced, integrated research agenda. It is clear, however, that the breadth and intensity of the agenda will depend on the resources available. Nonetheless, the interdependence of the many components of this research Plan is critical and the final approach needs to maintain balance among the various research foci, within the resources that are available. Greater commitment of resources will allow more complete understanding sooner. We believe that the importance of carbon cycle research within the pressures of confronting global change calls for an accelerated commitment of resources and that the Plan outlined here can be implemented efficiently and effectively.

The current state of carbon cycle science, together with the four research elements identified in Chapter 5, suggests near-term priorities for activities and programs to reach the stated goals. Our ability to reach these goals within the next decade, however, depends strongly on the ability of the U.S. funding agencies to provide full financial support to the research agenda. The 2009 *Our Changing Planet, an annual report to the*

U.S. Congress estimated that the total U.S. carbon cycle science budget is currently about $170 million per year (US Global Change Research Program, 2009a). Note that this estimate does not include the full cost of leveraged assets such as satellites, ship, or aircraft time. The current science budget is more than double the funding level at the time that the 1999 Science Plan was published, ignoring inflation, but most of these increases have just come in the past couple of years; the numbers still fall short of the $200 to $250 million per year budget recommended in the 1999 Science Plan. A summary of the element budgets estimated in Chapter 5 suggests that the total level of support for the new Plan needs to be approximately $500 million per year to expand and broaden the scope of carbon cycle research and to meet all of the stated goals. We believe that the critical, current importance of understanding global change is fully suggestive of support at this level.

Although the importance of understanding the global carbon cycle has been repeated in numerous planning documents and is receiving increased attention from the U.S. Congress, there are many factors that ultimately determine the annual carbon cycle research budgets. Evaluating and balancing priorities and determining what is achievable with only partial funding is difficult. Here we present three investment scenarios to provide some concept of what might be possible within 10 years with different funding levels. Each scenario is described in terms of its funding for program elements and its expected outcomes are characterized in terms of meeting the goals of this Plan. Lower levels of funding would limit the range of research activities and/or delay the accomplishment of some of the stated goals.

7.3.1 Scenario I: Full investment in carbon cycle research and observations (~$500M/yr)

Priorities for a full, integrated carbon research agenda are described throughout this Plan. Observational networks will be constructed to levels adequate for detecting and attributing change. Data management, synthesis, and modeling tools will be developed to take advantage of these observations. A coordinated information service will be developed to provide and disseminate carbon cycle information and products that are easily digestible by the general public, that are delivered in real time in some cases, and that take into account the needs of decision makers from beginning to end.

Expected outcomes for each of the goals outlined in Chapter 4:

- *Goal 1:* Full carbon observing networks, including of human systems, operating along with a plan for ensuring data quality, continuity, management, and access; advanced models that include the latest process-based understanding of carbon flux variability; and mechanistic understanding of responses and feedbacks to changing greenhouse gas concentrations and climate based on manipulative experiments and process studies.

- *Goal 2:* The relative importance of various socioeconomic processes and their interactions in different parts of the world and at a range of spatial and temporal scales are quantified; the potential range of future emissions from energy and land use are quantified; studies are published showing how carbon prices, institutions, and other policies affect socioeconomic drivers and emissions; and an integrated suite of tools, observations, and models is available for quantifying and evaluating emissions.

- *Goal 3:* Vulnerable pools and flows identified and monitored, especially those that may change rapidly in the near future; the physical, chemical, and biological processes important in determining the degree of vulnerability of carbon pools and flows built into diagnostic and mechanistic models; predictions published on the likelihood, timing and extent of potential changes in vulnerable carbon stocks and flows based on numerical models and empirical methods; predictions published on the consequences of carbon management and sequestration schemes on vulnerable pools; and carbon management goals supported by scientists helping to prioritize the most vulnerable stocks and flows that require management and the resources that are needed.

- *Goal 4:* Fully developed research program on the responses of ecosystem productivity, biodiversity, and sustainability to changing levels of carbon dioxide and other greenhouse gases; published studies on the synergistic effects of rising CO_2 on ecosystems in the presence of altered patterns of climate and associated changes in weather, hydrology, sea level, and ocean circulation; and fully developed, sustained and integrated measurement network for ecosystems in support of scientific research as well as management and decision making.

- *Goal 5:* Mechanisms developed for evaluating, integrating, and balancing interconnected and potentially competing management goals within the context of carbon cycle science; published studies on the impacts of carbon management and sequestration strategies on sustainability of ecosystems, ecosystem services, and economic and social systems – including water resources, biodiversity, and human livelihoods and well-being; and the net climate effects of carbon management pathways, including CO_2 capture and storage (CCS) as well as albedo and other energy-balance components that influence temperature, are quantified.

- *Goal 6:* The fundamental dynamics of decision making as they affect large-scale trends and patterns in carbon stocks and flows are quantified; and decision maker needs for carbon cycle science information and for understanding and dealing with uncertainty are systematically addressed so they can begin to incorporate carbon-related factors into their decision making.

7.3.2   *Scenario II: Partial investment in expanded priorities (~$300M/yr)*

Priorities for a partial funding of the set of research priorities outlined in this Plan include a limited expansion of existing atmospheric, terrestrial, oceanic, and space-based observations, and funding toward integrating the social sciences and natural sciences. However, investments in coordination of programs would need to be limited and fewer opportunities could be created to elucidate the connections between the natural, physical, and social science feedbacks of the carbon cycle. Intensive process studies would be conducted to improve understanding of carbon drivers, but their scope would need to be limited. Development of new tools to model and synthesize the observations and process information would improve the utility of information, but again the scope of these efforts would be limited relative to the recommended funding scenario (Section 7.3.1). Outreach would continue, but coordination between outreach efforts would be limited.

Expected outcomes for each of the goals outlined in Chapter 4:

- *Goal 1:* A limited expansion of carbon observing networks implemented with an uncertain plan for continuity of key data streams and maximizing data

quality; models capable of constraining process-based understanding of carbon flux variability under development; and limited manipulative experiments conducted to provide mechanistic understanding of responses and feedbacks to changing greenhouse gas concentrations and climate.

- *Goal 2:* Initial studies published on the relative importance of different socioeconomic processes; initial studies published on how carbon prices and other policies affect socioeconomic drivers and emissions; and initial development of the tools, observations, and models needed to quantify and evaluate emissions.

- *Goal 3:* A preliminary listing of the potential magnitude and likelihood of the risk for vulnerable pools and flows; studies published on the physical, chemical, and biological processes important in determining the degree of vulnerability of carbon pools and flows, but not fully incorporated into models; and carbon management goals supported with publications helping to prioritize the most vulnerable stocks and flows that require management and the resources that are needed.

- *Goal 4:* Ecosystems at risk from ocean acidification, land use change, and other carbon cycle drivers and consequences identified; extensive studies published on the responses of ecosystem productivity, biodiversity, and sustainability to changing levels of carbon dioxide and other greenhouse gases; and limited enhanced capabilities for sustained and integrated observations of ecosystems in support of scientific research as well as management and decision making.

- *Goal 5:* Peer-reviewed papers on the likelihood of success and potential for feedback or tradeoffs on physical and human systems of a few specific proposed carbon management pathways.

- *Goal 6:* Decision maker needs addressed in publications based on available carbon cycle science information as decision makers begin to incorporate carbon-related factors into their work.

Overall, partial funding of the expanded set of priorities outlined in this Plan would result in expanded and improved understanding of the global carbon cycle, but with a coarser resolution, a longer processing time to provide needed information to policymakers, and less integration across various aspects of the carbon cycle.

7.3.3 Scenario III: No increased funding to support expanded priorities (~$200M/yr)

Priorities for a very limited carbon research agenda include little more than maintaining the continuity of existing atmospheric, terrestrial, and oceanic time-series measurements and ongoing model development. Programs with a substantial amount of risk or high costs are unlikely to be funded. There will likely be little improvement in the integrative aspects of carbon cycle research or in the spatial resolution of data products that would provide decision makers with critical information on climate change.

Expected outcomes for each of the goals outlined in Chapter 4:

- *Goal 1:* Continue ongoing observations with continuing concern about the continuity of key data systems; and maintain current models of process-based understanding of carbon flux variability.

- *Goal 2:* Begin to develop the tools, observations, and models needed to quantify and evaluate emissions; physical and social sciences only weakly integrated.

- *Goal 3:* Publications hypothesizing about vulnerable pools and flows; and studies initiated on the physical, chemical, and biological processes important in determining the degree of vulnerability of carbon pools and flows.

- *Goal 4:* Identification of primary ecosystems at risk from ocean acidification, land use change, and other carbon cycle drivers and consequences; and initial studies published on the responses of ecosystem productivity, biodiversity, and sustainability to changing levels of carbon dioxide and other greenhouse gases.

- *Goal 5:* Studies published that extrapolate observations to comment on a few specific proposed carbon management pathways.

- *Goal 6:* Limited communication with decision makers as they begin to incorporate carbon-related factors into their decision making.

Under the limited investment scenario, the agencies and carbon cycle research community will have to work together to find cost saving opportunities and ways to widen the incorporation of social science research, to maximize efficiency and interdisciplinary cooperation, and to maintain U.S. leadership in this critical research realm.

References

Amiro, B.D., A.G. Barr, T.A. Black, R. Bracho, M. Brown, J. Chen, K.L. Clark, K.J. Davis, A. Desai, S. Dore, V. Engel, J.D. Fuentes, A.H. Goldstein, M.L. Goulden, T.E. Kolb, M.B. Lavigne, B.E. Law, H.A. Margolis, T. Martin, J.H. McCaughey, L. Misson, M. Montes-Helu, A. Noormets, J.T. Randerson, G. Starr, and J. Xiao, 2010. Ecosystem carbon dioxide fluxes after disturbance in forests of North America. *Journal of Geophysical Research*, 115:G00K02, doi:10.1029/2010JG001390.

Arndt, D.S., M.O. Baringer, and M.R. Johnson, 2010. State of the Climate in 2009. *Bulletin of the American Meteorological Society*, 91:s1–s222. doi:10.1175/BAMS-91-7-StateoftheClimate.

Bates, N.R., 2007. Interannual variability of the oceanic CO_2 sink in the subtropical gyre of the North Atlantic Ocean over the last 2 decades. *Journal of Geophysical Research*, 112(C9): C09013, doi:10.1029/2006JC003759.

Bates, N.R., J.T. Mathis, and L.W. Cooper, 2009. Ocean acidification and biologically induced seasonality of carbonate mineral saturation states in the western Arctic Ocean. *Journal of Geophysical Research*, 114:C11007, doi:10.1029/2008JC004862.

Behrenfeld, M.J., R.T. O'Malley, D.A. Siegel, C.R. McClain, J.L. Sarmiento, G.C. Feldman, A.J. Milligan, P.G. Falkowski, R.M. Letelier, and E.S. Boss, 2006. Climate-driven trends in contemporary ocean productivity. *Nature*, 444:752–755.

Birdsey, R., N. Bates, M. Behrenfeld, K. Davis, S.C. Doney, R. Feely, D. Hansell, L. Heath, E. Kasischke, H. Kheshgi, B. Law, C. Lee, A.D. McGuire, P. Raymond, and C.J. Tucke, 2009. Carbon cycle observations: gaps threaten climate mitigation strategies. *Eos*, 90(34):292.

Brewer, P.G., and E.T. Peltzer, 2009. Limits to marine life. *Science*, 324:347–348.

Buesseler, K.O., C.H. Lamborg, P.W. Boyd, P.J. Lam, T.W. Trull, R.R. Bidigare, J.K.B. Bishop, K.L. Casciotti, F. Dehairs, M. Elskens, M. Honda, D.M. Karl, D.A. Siegel, M.W. Silver, D.K. Steinberg, J. Valdes, B. Van Mooy, and S. Wilson, 2007. Disappearing in the twilight zone: Revisiting carbon flux through the ocean's twilight zone. *Science*, 316(5824):567–570.

Buizer, J., K. Jacobs, and D. Cash, 2010. Making short-term climate forecasts useful: Linking science and Action. *Proceedings of the National Academy of Science*, doi:10.1073/pnas.0900518107.

Cai, W.-J., 2011. Estuarine and coastal ocean carbon paradox: CO_2 sinks or sites of terrestrial carbon incineration? *Annual Review of Marine Science*, 3:5.1–5.23, doi:10.1146/annurev-marine-120709-142723.

Cash, D.W., J.C. Borck, and A.G. Patt, 2006. Countering the loading-dock approach to linking science and decision making. Comparative analysis of El Niño/Southern Oscillation (ENSO) forecasting systems. *Science, Technology and Human Values*, 31:465-494.

Ciais, P., A.J. Dolman, R. Dargaville, L. Barrie, A. Bombelli, J. Butler, P. Canadell, and T. Moriyama, 2010. *GeoCarbon Strategy*. Geo Secretariat Geneva, FAO, Rome, 48 pp.

Climate Change Science Program and Subcommittee on Global Change Research, 2003. *Strategic Plan for the U.S. Climate Science Program*. Washington, DC, 364 pp.

Climate Change Science Program and Subcommittee on Global Change Research, 2008. *Revised Research Plan for the U.S. Climate Change Science Program*. Washington, DC, 78 pp.

Cox, P.M., R.A. Betts, C.D. Jones, S.A. Spall, and I.J. Totterdell, 2000. Acceleration of global warming due to carbon-cycle feedbacks in a coupled climate model. *Nature*, 408:184–187.

Denman, K.L., G. Brasseur, A. Chidthaisong, P. Ciais, P.M. Cox, R.E. Dickinson, D. Hauglustaine, C. Heinze, E. Holland, D. Jacob, U. Lohmann, S. Ramachandran, P.L. da Silva Dias, S.C. Wofsy, and X. Zhang, 2007. Coupling Between Changes in the Climate System and Biogeochemistry. In: *Climate Change 2007: The Physical Science Basis. Contribution of Working Group I to the Fourth Assessment Report of the Intergovernmental Panel on Climate Change.* Cambridge University Press, Cambridge, United Kingdom and New York, NY, USA.

Denning, A.S., et al., 2005. *Science Implementation Strategy for the North American Carbon Program. Report of the NACP Implementation Strategy Group of the U.S. Carbon Cycle Interagency Working Group.* U.S. Carbon Cycle Science Program, Washington, DC.

Dilling, L., 2007. Towards science in support of decision making: characterizing the supply of carbon cycle science. *Environmental Science and Policy*, 10:48–60.

Dilling, L., and M.C. Lemos, 2011. Defining usable science: What can we learn for science policy from the seasonal climate forecasting experience? *Global Environmental Change*, doi:10.1016/j.gloenvcha.2010.11.006.

Doney, S.C., R. Anderson, J. Bishop, K. Caldeira, C. Carlson, M.-E. Carr, R. Feely, M. Hood, C. Hopkinson, R. Jahnke, D. Karl, J. Kleypas, C. Lee, R. Letelier, C. McClain, C. Sabine, J. Sarmiento, B. Stephens, and R. Weller, 2004. *Ocean Carbon and Climate Change (OCCC):An Implementation Strategy for U.S. Ocean Carbon Cycle Science.* University Center for Atmospheric Research, Boulder, CO, 108 pp.

Doney, S.C., V.J. Fabry, R.A. Feely, and J.A. Kleypas, 2009. Ocean acidification: the other CO_2 problem. *Annual Review of Marine Science*, 1:169–192.

Donner, R., S. Barbosa, J. Kurths, and N. Marwan (eds.), 2009. Understanding the Earth as a complex system: Advances in data analysis and modeling in earth sciences. *European Physical Journal Special Topics*, 174:1–9.

Duarte, C.M., J.J. Middelburg, and N. Caraco, 2005. Major role of marine vegetation on the oceanic carbon cycle. *Biogeosciences*, 2:1–8.

Enting, I.G., 2008. Seeking carbon-consistency in the climate-science-to-policy interface. *Biogeochemistry*, doi:10.1007/s10533-009-9351-7.

Fabry, V.J., C. Langdon, W.M. Balch, A.G. Dickson, R.A. Feely, B. Hales, D.A. Hutchins, J.A. Kleypas, and C.L. Sabine, 2008. *Present and Future Impacts of Ocean Acidification on Marine Ecosystems and biogeochemical cycles.* University of California at San Diego, Scripps Institution of Oceanography, Woods Hole, MA.

Failey, E. and L. Dilling, 2010. Carbon stewardship: Land management decisions and the potential for carbon sequestration in Colorado, USA. *Environmental Research Letters*, 5, doi:10.1088/1748-9326/5/2/024005.

Fargione, J., J. Hill, D. Tilman, S. Polasky, and P. Hawthorne, 2008. Land clearing and the biofuel carbon debt. *Science*, 319:1235–1238.

Field, C., D.B. Lobell, H.A. Peters, and N. Chiariello, 2007. Feedbacks of terrestrial ecosystems to climate change. *Annual Review of Environment and Resources*, doi:10.1146/annurev.energy.1132.053006.141119.

Friedlingstein, P., P. Cox, R. Betts, L. Bopp, W. von Bloh, V. Brovkin, P. Cadule, S. Doney, M. Eby, I. Fung, G. Bala, J. John, C. Jones, F. Joos, T. Kato, M. Kawamiya, W. Knorr, K. Lindsay, H.D. Matthews, T. Raddatz, P. Rayner, C. Reick, E. Roeckner, K.-G. Schnitzler, R. Schnur, K. Strassmann, A.J. Weaver, C. Yoshikawa, and N. Zeng, 2006. Climate-carbon cycle feedback analysis: Results from the C4MIP Model Intercomparison. *Journal of Climate*, 19:3337–3353.

GCP, 2010. *Global Carbon Project.* At www.globalcarbonproject.org, accessed 9 Nov. 2010.

GLP, 2010. *Global Land Project.* At www.globalandproject.org, accessed 9 Nov. 2010.

Gill, R.A., H.W. Polley, H.B. Johnson, L.J. Anderson, H. Maherali, and R.B. Jackson, 2002. Nonlinear grassland responses to past and future atmospheric CO_2. *Nature*, 417:279–282.

Houghton, R.A., 2007. Balancing the global carbon budget. *Annual Review of Earth and Planetary Sciences*, 35:313–347.

Hudiburg, T., B. Law, D.P. Turner, J. Campbell, D. Donato, and M. Duane, 2009. Carbon dynamics of Oregon and Northern California forests and potential land-based carbon storage. *Ecological Applications*, 19(1):163–180.

IIASA, 2010. *Global Energy Assessment*. International Institute for Applied Systems Analysis, Laxenburg, Austria. At www.iiasa.ac.at/Research/ENE/GEA/index_gea.html, accessed 9 Nov. 2010.

IPCC, 2006. *2006 IPCC guidelines for National Greenhouse Gas Inventories. Vol 4. Agriculture, forestry and other land use (AFOLU)*. Prepared by the National Greenhouse Gas Inventories Programme. Eggleston, H.S., L. Buendia, K. Miwa, T. Ngara, and K. Tanabe (eds.). Institute for Global Environmental Strategies, Hayama, Japan. Available at: www.ipcc-nggip.iges.or.jp/public/2006gl/vol4.html.

IPCC, 2007. *Climate Change 2007 Synthesis Report. Contribution of Working Groups I, II and III to the Fourth Assessment Report of the Intergovernmental Panel on Climate Change*. Core Writing Team, R.K. Pachauri, and A. Reisinger, A. (eds.). IPCC, Geneva, Switzerland, 104 pp.

IPCC, 2011. *Technical Papers*. At www.ipcc.ch/publications_and_data/publications_and_data_technical_papers.shtml, accessed May 2011.

Jackson, R.B., E.G. Jobbágy, R. Avissar, S. Baidya Roy, D. Barrett, C.W. Cook, K.A. Farley, D.C. le Maitre, B.A. McCarl, and B. Murray, 2005. Trading water for carbon with biological carbon sequestration. *Science*, 310:1944–1947.

Jones, C., J. Lowe, S. Liddicoat, and R. Betts, 2009. Committed terrestrial ecosystem changes due to climate change. *Nature Geoscience*, 2:484 –487.

Kashian, D.M., W.H. Romme, D.B. Tinker, M.G. Turner, and M.G. Ryan, 2006. Carbon storage on landscapes with stand-replacing fires. *BioScience*, 56(7):598–606.

Körner, C., 2009. Responses of humid tropical trees to rising CO_2. *Annual Review of Ecology, Evolution, and Systematics*, 40:61–79

Lambin, E.F., et al., 2001. The causes of land-use and land-cover change: Moving beyond the myths. *Global Environmental Change: Human and Policy Dimensions*,11(4): 261–269.

Law, B.E., and M.E. Harmon, 2011. Forest sector carbon management, measurement and verification, and discussion of policy related to climate change. *Carbon Management*, 2(1):73–84.

Lemos, M.C., and Morehouse, B.J., 2005. The co-production of science and policy in integrated climate assessments. *Global Environmental Change*, 15:57-68.

Logar, N., Conant, R, 2007. Reconciling the Supply of and Demand for Carbon Cycle Science in the U.S. Agricultural Sector. *Environmental Science and Policy* 10:1(75–84).

Lowrey, J., A. Ray, and R. Webb, 2009. Factors influencing the use of climate information by Colorado municipal water managers. *Climate Research*, 40:103–119.

Millennium Ecosystem Assessment, 2003. *Ecosystems and Human Well-being: A Framework for Assessment*. Island Press.

Morgan, M.G., H. Dowlatabadi, M. Henrion, D. Keith, R. Lempert, S. McBride, M. Small, and T. Wilbanks, 2009. *Best Practice Approaches for Characterizing, Communicating, and Incorporating Scientific Uncertainty in Climate Decision Making. Synthesis and Assessment Product 5.2*. Report by the U.S. Climate Change Science Program and the Subcommittee on Global Change Research, National Oceanic and Atmospheric Administration, Washington, DC, 96 pp.

Morss, R., O. Wilhelmi, M. Downton, and E. Gruntfest, 2005. Lessons from an interdisciplinary project. *Bull. Am. Met. Soc.*, doi:10.1175/BAMS-86-11-1593.

Morss, R.E., J.L. Demuth, J.K. Lazo, 2008. Communicating uncertainty in weather forecasts: A survey of the U.S. public. *Weather and Forecasting*, 23(5):974–991.

Moser, S., 2009. Making a difference on the ground: the challenge of demonstrating the effectiveness of decision support. *Climatic Change*, 95:11 –21.

Moser, S., and L. Dilling (eds.), 2007. *Creating a Climate for Change: Communicating Climate Change—Facilitating Social Change*. Cambridge University Press, Cambridge, UK.

NACP, 2010a. *North American Carbon Program*. At www.nacarbon.org/nacp/index.html, accessed Nov. 2010.

NACP, 2010b. *Modeling and Synthesis Thematic Data Center (MAST-DC)*. nacp.ornl.gov/, accessed 4 Nov. 2010.

National Research Council, 2007a. *Earth Science and Applications from Space: National Imperatives for the Next Decade and Beyond*. The National Academies Press, Washington, DC, 456 pp.

National Research Council, 2007b. *Evaluating Progress of the U.S. Climate Change Science Program: Methods and Preliminary Results*. National Academies Press, Washington, DC, 170 pp.

National Research Council, 2009a. *Informing Decisions in a Changing Climate—Panel on Strategies and Methods for Climate-Related Decision Support*. National Research Council of the National Academies, Washington, DC.

National Research Council, 2009b. *Restructuring Federal Climate Research to Meet the Challenges of Climate Change*. National Academies Press, Washington, DC.

National Research Council, 2010a. *Advancing the Science of Climate Change*. National Academies Press, Washington, DC, 506 pp.

National Research Council, 2010b. *Limiting the Magnitude of Future Climate Change*. National Academies Press, Washington, DC, 258 pp.

National Research Council, 2010c. *Adapting to the Impacts of Climate Change*. National Academies Press, Washington, DC, 325 pp.

National Research Council, 2010d. *Verifying Greenhouse Gas Emissions: Methods to Support International Climate Agreements*. National Academies Press, Washington, DC.

National Research Council, 2010e. *Ocean Acidification: A National Strategy to Meet the Challenges of a Changing Ocean*. National Academies Press, Washington, DC.

NEON, 2010. *National Ecological Observatory Network*. At www.neoninc.org, accessed 4 Nov. 2010.

NOAA, 2010. *National Oceanographic Data Center*. At www.nodc.noaa.gov, accessed 4 Nov. 2010. National Oceanic and Atmospheric Administration.

ORNL, 2010. *Ameriflux*. At public.ornl.gov/ameriflux, accessed 4 Nov. 2010. Oak Ridge National Laboratory.

Page, S.E., F. Siegert, J.O. Rieley, H.-D.V. Boehm, A. Jaya, and S. Limin, 2002. The amount of carbon released from peat and forest fires in Indonesia during 1997. *Nature*, 420:61–65.

Piñeiro, G., E.G. Jobbágy, J. Baker, B.C. Murray, and R.B. Jackson, 2009. Set-asides can be better climate investment than corn-ethanol. *Ecological Applications*, 19:277–282.

Pörtner, H., 2008. Ecosystem effects of ocean acidification in times of ocean warming: a physiologist's view. *Mar. Ecol. Prog. Ser.*, 373:203–217.

Potter, C., 2010. The carbon budget of California. *Environmental Science and Policy*, 13:373–383.

Qian, H.-F., R. Joseph, and N. Zeng, 2010. Enhanced terrestrial carbon uptake in the northern high latitudes in the 21st century from the C4MIP model projections. *Global Change Biology*, doi:10.1111/j.1365-2486.2009.01989.

Raupach, M., and J. Canadell, 2007. Observing a vulnerable carbon cycle. In: *Observing the Continental Scale Greenhouse Gas Balance of Europe*. H. Dolman, R. Valentini, and A. Friebauer (eds.). Springer.

Raupach, M. et al., 2007. Global and regional drivers of accelerating CO_2 emissions. *Proceedings of the National Academy of Sciences*, 104:10288–10293.

Richards, K.R., R.N. Sampson, and S. Brown, 2006. *Agricultural and forestlands: US carbon Policy Strategies*. A report prepared for the Pew Center on Global Climate Change. Available at: www.pewclimate.org/reports/All (accessed 15 Mar. 2010).

Riebesell, U., 2008. Climate change: Acid test for marine biodiversity. *Nature*, 454:46–47.

Root, T.L., J.T. Price, K. Hall, S.H. Schneider, C. Rosenzweig, and J.A. Pounds, 2004. Fingerprints of global warming on wild animals and plants. *Nature*, 421:57 –60.

Rosenzweig, C., D. Karoly, M. Vicarelli, P. Neofotis, Q. Wu, G. Casassa, A. Menzel, T.L. Root, N. Estrella, B. Seguin, P. Tryjanowski, C. Liu, S. Rawlins, and A. Imeson, 2008. Attributing physical and biological impacts to anthropogenic climate change. *Nature*, 453:353 –358.

Sarmiento, J.L., and S.C. Wofsy, 1999. *A U.S. Carbon Cycle Science Plan*. Report of the Carbon and Climate Working Group for the U.S. Global Change Research Program. U.S. Global Change Research Program, Washington, DC.

Schimel, D.S., J.I. House, K.A. Hibbard, P. Bousquet, P. Ciais, P. Peylin, B.H. Braswell, M.J. Apps, D. Baker, A. Bondeau, J. Canadell, G. Churkina, W. Cramer, A.S. Denning, C.B. Field, P. Friedlingstein, C. Goodale, M. Heimann, R.A. Houghton, J.M. Melillo, B. Moore, D. Murdiyarso, I. Noble, S.W. Pacala, I.C. Prentice, M.R. Raupach, P.J. Rayner, R.J. Scholes, W.L. Steffen, and C. Wirth, 2001. Recent patterns and mechanisms of carbon exchange by terrestrial ecosystems. *Nature*, 414:169–172.

Schneider, S.H., S. Semenov, A. Patwardhan, I. Burton, C.H.D. Magadza, M. Oppenheimer, A.B. Pittock, A. Rahman, J.B. Smith, A. Suarez, and F. Yamin, 2007. Assessing key vulnerabilities and the risk from climate change. In: *Climate Change 2007: Impacts, Adaptation and Vulnerability. Contribution of Working Group II to the Fourth Assessment Report of the Intergovernmental Panel on Climate Change*. M.L. Parry, O.F. Canziani, J.P. Palutikof, P.J. van der Linden and C.E. Hanson (eds.). Cambridge University Press, Cambridge, UK, pp. 779–810.

Schuur, E.A.G., J. Bockheim, J.G. Canadell, E. Euskirchen, C.B. Field, S.V. Goryachkin, S. Hagemann, P. Kuhry, P.M. Lafleur, H. Lee, G. Mazhitova, F.E. Nelson, A. Rinke, V.E. Romanovsky, N. Shiklomanov, C. Tarnocai, S. Venevsky, J.G. Vogel, and S.A. Zimov, 2008. Vulnerability of permafrost carbon to climate change: Implications for the global carbon cycle. *BioScience*, 58(8):701–714, doi:10.1641/B580807.

Schwalm, C.R., C.A. Williams, K. Schaefer, R. Anderson, M.A. Arain, I. Baker, A. Barr, T.A. Black, G. Chen, J.M. Chen, P. Ciais, K.J. Davis, A. Desai, M. Dietze, D. Dragoni, M.L. Fischer, L.B. Flanagan, R. Grant, L. Gu, D. Hollinger, R.C. Izaurralde, C. Kucharik, P. Lafleur, B.E. Law, L. Li, Z. Li, S. Liu, E. Lokupitiya, Y. Luo, S. Ma, H. Margolis, R. Matamala, H. McCaughey, R.K. Monson, W.C. Oechel, C. Peng, B. Poulter, D.T. Price, D.M. Riciutto, W. Riley, A.K. Sahoo, M. Sprintsin, J. Sun, H. Tian, C. Tonitto, H. Verbeeck, and S.B. Verma, 2010. A model-data intercomparison of CO_2 exchange across North America: Results from the North American Carbon Program site synthesis. *Journal of Geophysical Research – Biogeosciences*, 115, G00H05, doi:10.1029/2009JG001229.

Searchinger, T., R. Heimlich, R.A. Houghton, F. Dong, A. Elobeid, J. Fabiosa, S. Tokgoz, D. Hayes, and T.-H. Yu, 2008. Use of U.S. croplands for biofuels increases greenhouse gases through emissions from land-use change. *Science*, 319:1238–1240.

SOCCR, 2007. *The first State of the Climate Cycle Report (SOCCR): The North American Carbon Budget and Implications for the Global Carbon Cycle*. A.W. King, L. Dilling, G.P. Zimmerman, D.M. Fairman, R.A. Houghton, G. Marland, A.Z. Rose, and T.J. Wilbanks (eds.). Climate Change Science Program and Subcommittee on Global Change Research, National Oceanic and Atmospheric Administration, Asheville, NC, 242 pp.

Stern, P.C. 2002. *Human Interactions with the Carbon Cycle: Summary of a Workshop*. Committee on the Human Dimensions of Global Change. National Academies Press, Washington, DC.

Tschakert, P., E. Huber-Sannwald, D.S. Ojima, M.R. Raupach, and E. Schienke, 2008. Principles for holistic, adaptive management of the terrestrial carbon cycle at local and regional scales, through multi-criteria analysis. *Global Environmental Change*, 18:128–141.

Turner, W.R., M. Oppenheimer, and D.S. Wilcove, 2009. A force to fight global warming. *Nature*, 462:278–279.

UGEC, 2010. *Urbanization and Global Environmental Change*. At www.ugec2010.org, accessed 9 Nov. 2010.

US DOE, 2007. *Scenarios of greenhouse gas emissions and atmospheric concentrations (Part A) and review of integrated scenario development and application (Part B)*. A report by the U.S. Climate Change Science Program and the Subcommittee on Global Change Research (Clarke, L., J. Edmonds, J. Jacoby, H. Pitcher, J. Reilly, R. Richels, E. Parson, V. Burkett, K. Fisher-Vanden, D. Keith, L. Mearns, C. Rosenzweig, and M. Webster (authors)), U.S. Department of Energy, Office of Biological and Environmental Research, Washington DC, 260 pp.

U.S. Global Change Research Program 1989. *Our Changing Planet: A U.S. Strategy for Global Change Research*, Washington, D.C., 38 pp.

U.S. Global Change Research Program, 2009a. *Our Changing Planet, A report by the U.S. Global Change Research Program and the Subcommittee on Global Change Research: A supplement to the President's budget for fiscal year 2010*, USGCRP, Washington D.C., 172pp.

U.S. Global Change Research Program 2009b. *Global Climate Change Impacts in the United States.*

USDA Forest Service, 2010. *Forest Inventory and Analysis National Program*. At http://fia.fs.fed.us/, accessed 4 Nov. 2010. U.S. Department of Agriculture, Forest Service.

Walter, K.M., S. Zimov, J.P. Chanton, D. Verbyla, and F.S. Chapin III, 2006. Methane bubbling from Siberian thaw lakes as a positive feedback to climate warming. *Nature*, 443:71–75.

Waycott, M., C.M. Duarte, T.J.B. Carruthers, R.J. Orth, W.C. Dennison, S. Olyarnik, A. Calladine, J.W. Fourqurean, K.L. Heck, A.R. Hughes, G.A. Kendrick, W.J. Kenworthy, F.T. Short, and S. L. Williams, 2009. Accelerating loss of seagrasses across the globe threatens coastal ecosystems. *Proceedings of the National Academy of Sciences*, 106(30):12377–12381.

Williams, J.W., S.T. Jackson, and J.E. Kutzbach, 2007. Projected disturbances of novel and disappearing climates by 2100. *Proceedings of the National Academy of Sciences*, 104:5738–5742.

Wofsy, S.C., and R.C. Hariss, 2002. *The North American Carbon Program Plan (NACP). Report of the NACP Committee of the U.S. Carbon Cycle Science Program*. U.S. Global Change Research Program, Washington, DC.

Appendix A:

Charge to the Co-Leads of the Carbon Cycle Science Working Group, and Overview of the Carbon Cycle Interagency Working Group

A.1 Charge to the co-leads of the Carbon Cycle Science Working Group

A New U.S. Carbon Cycle Science Plan
(CCIWG Approved 18 May 2008)

Rationale: *A U.S. Carbon Cycle Science Plan* (Sarmiento and Wofsy, 1999) was developed in 1998, published in 1999, and is now essentially 10 years old. Much has been learned and there is no doubt much yet to be done, but it is time to take a fresh look at the scientific questions and priorities detailed in that report. It is important to note that this plan, produced by the scientific community, was the single most important and influential input into the Carbon Cycle chapter of the 2003 *Strategic Plan for the U.S. Climate Change Science Program*. The U.S. Climate Change Science Program (CCSP) is now working on a minor update of its 2003 Strategic Plan and intends to draft a major revision in 2009. CCSP leaders have asked the Carbon Cycle Interagency Working Group (CCIWG) to identify by December 2008 the "building blocks" it will use to develop its contribution to the revised strategic plan. The CCIWG would again like to have an up-to-date report from the scientific community on the most important scientific challenges and priorities for U.S. carbon cycle research as the major "building block" to draw upon in drafting its inputs for the new strategic plan. If initiated immediately, there would be time to complete a community-based study similar to the one produced by the Carbon and Climate Working Group led by Jorge Sarmiento and Steve Wofsy in 1998-1999.

Immediate Actions to Initiate Planning: The CCIWG should consult with its Carbon Cycle Science Steering Group (CCSSG) to request their assistance in defining and organizing the community-based planning activity needed to develop the new report. It does not seem reasonable to charge the CCSSG with developing the report itself (although individual members may wish to participate), but rather they should help to define the process and identify the working group participants.

It would be reasonable to make the new working group a subcommittee of the CCSSG, if the CCSSG agrees. The working group will need to develop a work plan and cost plan that can guide their activities and schedule and serve to justify the resources to be provided through the CCIWG agencies. The working group will be responsible for end-to-end implementation of the planning process, but it is anticipated that some of the authors for the final report, perhaps even the lead authors or editors, may emerge through leadership roles assumed by other community members as the planning proceeds. Working group members should represent the composition of the community as well as possible and include active researchers likely to be engaged in the next 10 years of carbon cycle research.

Charge to the New Working Group: The carbon cycle science working group will be responsible for developing an updated, revised, or new science plan for U.S. carbon cycle science, identifying challenges and priorities for the next decade (~2010-2020) and involving the broader community. The group will:

- Define a process that reaches out to and engages the U.S. carbon cycle science community at key stages (for example, one or more community workshops and inviting many to participate in a peer review of the report), but that is no more elaborate and lengthy than is needed to do the job.

- Consider how to engage other key stakeholders to ensure that their interests and priorities are taken into account – especially key decision-support needs.

- Develop a schedule that as much as possible matches CCSP planning needs (for example, to have preliminary findings available as near to December 2008 as possible and final publication of the report before content is fixed for the next strategic plan for the U.S. CCSP)

- Prepare a work plan and cost plan for the planning activities and report preparation and submit this to the CCIWG agencies for internal review, approval and funding.

- Implement the planning process, holding whatever meetings and workshops are included in the approved work plan, to identify the most important science issues for the U.S. to pursue and what is needed to address them. The following should be taken into consideration (the order below is not a prioritization):

  - The most important, exciting, challenging science questions that are ripe for new investment.

  - The U.S. Government's need for prioritized research to address critical uncertainties regarding global climate change.

  - The most important observations and research infrastructure in need of continuing, stable support.

  - The needs of policy makers and resources managers for decision support related to carbon management, climate change mitigation (including emissions reduction and carbon sequestration) and/or adaptation.

  - The previous (1999) report on U.S. carbon cycle science – what is no longer important, what needs only updating, what requires major revision, what needs to be added? Use it as a starting point, if possible, but if something wholly new would be best, that would be acceptable.

  - Existing carbon cycle science budget levels and anticipated future funding levels; recommended activities and priorities should either be more or less affordable within existing budgets or tied to well-defined initiatives that could be proposed with high priority for new funding (Note: both should be included!)

  - The missions/goals of the U.S. agencies that conduct carbon cycle science and the relevant scientific and/or operational infrastructure that they are mandated to support.

  - International programs, plans, priorities for carbon cycle science

- Write and publish a report on the findings of the working group.

  - Select the editor(s) for the report (suggest co-editors to cover the span of "disciplinary expertise needed – at least land-ocean, perhaps land-atmosphere-ocean?), subject to the concurrence of the CCSSG Chair and the CCIWG.

  - Recruit additional authors outside of the working group, as needed.

  - Keep the CCIWG apprised of findings and status of the report and seek their comments/inputs at an appropriate time(s)

  - Make a mature draft of the report available for community review and comment, and revise the draft to appropriately respond to this review

  - Arrange for the final report to be made available (help with electronic posting and/or printing could be arranged through the Carbon Cycle Science Program Office and UCAR)

A.2 Carbon Cycle Interagency Working Group

The Carbon Cycle Interagency Working Group was established under the U.S. Climate Change Science Program to promote interagency cooperation and coordination, help secure funding, prepare individual and joint agency initiatives and solicitations, and involve the scientific community with the purpose of providing the needed science to understand the carbon cycle. CCIWG members represent 12 federal agencies.

- Department of Agriculture
  - Agricultural Research Service
  - Economic Research Service
  - Forest Service
  - National Institute of Food and Agriculture
  - Natural Resource Conservation Service
- Department of Energy
- Environmental Protection Agency
- National Aeronautics and Space Administration
- National Institute of Standards and Technology
- National Oceanic and Atmospheric Administration, Department of Commerce
- National Science Foundation
- US Geological Survey, Department of the Interior

Appendix B:
Carbon Cycle Science Working Group Membership

Robert F. Anderson
Lamont-Doherty Earth Observatory
Earth and Environmental Sciences
The Earth Institute
Columbia University

Deborah Bronk
Department of Physical Sciences
Virginia Institute of Marine Sciences
College of William & Mary

Kenneth J. Davis
Department of Meteorology
Pennsylvania State University

Ruth S. DeFries
Department of Ecology, Evolution, and Environmental
    Biology
Columbia University

A. Scott Denning
Department of Atmospheric Science
Colorado State University

Lisa Dilling
Center for Science and Technology Policy Research
Cooperative Institute for Research in Environmental Sciences
University of Colorado, Boulder

Robert B. Jackson – Co-lead
Department of Biology
Nicholas School of the Environment
Duke University

Andy Jacobson
NOAA Earth System Research Laboratory, Global Monitoring
    Division
University of Colorado, Cooperative Institute for Research in
    Environmental Sciences

Steven Lohrenz
Department of Marine Science
University of Southern Mississippi

Gregg Marland – Co-lead
Ecosystem Simulation Science Group, Environmental Sciences
    Division
Oak Ridge National Laboratory
now at
Research Institute for Environment, Energy, and Economics
Appalachian State University

A. David McGuire
Alaska Cooperative Fish and Wildlife Research Unit
    U.S. Geological Survey
    Institute of Arctic Biology
Department of Biology and Wildlife
University of Alaska

Galen A. McKinley
Department of Atmospheric and Oceanic Sciences
University of Wisconsin, Madison

Anna M. Michalak – Co-lead
Department of Civil and Environmental Engineering
Department of Atmospheric, Oceanic and Space Sciences
University of Michigan
now at
Department of Global Ecology
The Carnegie Institution for Science

Charles Miller
Jet Propulsion Laboratory
California Institute of Technology

Berrien Moore III
College of Atmospheric and Geographic Sciences
University of Oklahoma

Dennis S. Ojima
Natural Resource Ecology Laboratory
Colorado State University
Heinz Center

Brian O'Neill
Integrated Assessment Modeling, Climate Change Research
National Center for Atmospheric Research

James T. Randerson
Department of Earth System Science
University of California, Irvine

Steven W. Running
Numerical Terradynamic Simulation Group
College of Forestry and Conservation
University of Montana

Christopher L. Sabine – Co-lead
Ocean Climate Research Division
NOAA Pacific Marine Environmental Laboratory

Brent Sohngen
Department of Agricultural, Environmental and Development
    Economics
Ohio State University

Pieter P. Tans
Climate Monitoring and Diagnostics Laboratory
NOAA Earth System Research Laboratory

Peter E. Thornton
Environmental Simulation Science Group
Oak Ridge National Laboratory

Steven C. Wofsy
School of Engineering and Applied Sciences
Harvard Forest
Harvard University

Ning Zeng
Department of Atmospheric and Oceanic Science
University of Maryland

Appendix C:
Outreach Activities

The following is a list of meetings, workshops, conferences, and publications where information about the new U.S. Carbon Cycle Science Plan has been presented and discussed:

November 17-18, 2008	Carbon Cycle Science Working Group (CCS WG) Meeting Washington, DC Scope: Dedicated workshop
December 9-10, 2008	Carbon Cycle Science Steering Group (CCSSG) Meeting Washington, DC Scope: Report on CCS WG meeting and presentation
February 17-20, 2009	North American Carbon Program (NACP) All Investigators' Meeting San Diego, CA Scope: CCS WG side meeting, plenary presentation, dedicated breakout session
March 24, 2009	"A U.S. carbon cycle science plan: First meeting of the Carbon Cycle Science Working Group; Washington, D.C., 17-18 November 2008" by A.M Michalak, R. Jackson, G. Marland, and C. Sabine, published in EOS, Transactions of the American Geophysical Union, 90(12), pp. 102-103.
March 27, 2009	CCS WG Scoping Paper published online at www.carboncyclescience.gov/carbonplanning.php
May 24-27, 2009	2009 Joint Assembly, The Meeting of the Americas Toronto, Ontario, Canada Scope: Presentation
June 1-2, 2009	CCS WG Meeting Washington, DC Scope: Dedicated workshop
June 3-4, 2009	CCSSG Meeting Washington, DC Scope: Report on CCS WG meeting and presentation
June 23-25, 2009	Earth System Science Partnership (ESSP) Global Carbon Project (GCP) Science Steering Committee (SSC) Beijing ,China Scope: Progress report and presentation
July 20-23, 2009	Ocean Carbon and Biogeochemistry (OCB) Summer Workshop Woods Hole, MA Scope: Presentations and panel discussion
August 2-7, 2009	94th Ecological Society of America Meeting Albuquerque, NM Scope: Presentation

September 13-19, 2009	8th International Carbon Dioxide Conference Jena, Germany Scope: Abstract and poster presentation
September 21-25, 2009	OceanObs'09 Venice-Lido, Italy Scope: Presentation and poster
September 21-23, 2009	AmeriFlux Meeting Washington, DC Scope: Presentation
September 30, 2009	NACP Science Steering Group (NACP SSG) Washington, DC Scope: Presentation
November 6, 2009	CCSWG. Recommendations Summary White Paper published online at www.carboncyclescience.gov/carbonplanning.php
November 19-20, 2009	39th National Research Council (NRC) Committee on Human Dimensions of Global Change Meeting Washington, DC Scope: Presentation
December 3-4, 2009	CCSSG Meeting Washington, DC Scope: Progress report and presentation
December 14-18, 2009	American Geophysical Union (AGU) Fall Meeting San Francisco, CA Scope: Town hall meeting
March 12-13, 2010	Human Dimensions Workshop Washington, DC Scope: Dedicated workshop
June 15-17, 2010	ESSP GCP SSC Meeting Norwich, United Kingdom Scope: Progress report and discussion
July 13-14, 2010	NASA Carbon Monitoring System Scoping Workshop Boulder, CO Scope: Presentation and discussion
July 19-22, 2010	OCB Summer Workshop San Diego, CA Scope: Presentation and discussion
August 23-24, 2010	CCS WG Workshop Boulder, CO Scope: Dedicated workshop

Appendix D:
List of Acronyms

ACE	Aerosol–Cloud–Ecosystems
AIRS	Atmospheric Infrared Sounder
ALOS	Advanced Land Observing Satellite
AmeriFlux	Tower network that provides continuous observations of ecosystem level exchanges of CO_2, water, energy and momentum, composed of sites from North, South, and Central America.
ASCENDS	Active Sensing of CO2 Emissions over Nights, Days, and Seasons
CCIWG	Carbon Cycle Interagency Working Group
CCS	carbon capture and storage
CCSP	Climate Change Science Program
CCSSG	Carbon Cycle Science Steering Group
CH_4	Methane
CO_2	Carbon dioxide
DOE	Department of Energy
EOS	Earth Observing System
ESRL	NOAA Earth System Research Laboratory
FACE	Free Air CO_2 Enrichment
FIA	USDA Forest Service Forest Inventory and Analysis
FLUXNET	Global network of micrometeorological tower sites that use eddy covariance methods to measure the exchanges of CO_2, water vapor, and energy between terrestrial ecosystems and the atmosphere.
GCM	General Circulation Model
GCP	Global Carbon Project
GEO	Group on Earth Observations
GEO-CAPE	Geostationary Coastal and Air Pollution Events
GOSAT	Greenhouse Gas Observing Satellite
GPP	Gross Primary Productivity
HyspIRI	Hyperspectral Infrared Imager
IASI	Infrared Atmospheric Sounding Interferometer
ICESat-2	Ice, Cloud, and land Elevation Satellite-2
ICSU	International Council for Science
IMBER	Integrated Marine Biogeochemistry and Ecosystem Research
IOOS	NOAA Integrated Ocean Observing System
IPCC	Intergovernmental Panel on Climate Change
JPSS	Joint Polar Satellite System
LDCM	Landsat Data Continuity Mission
LIDAR	Light Detection and Ranging
LIST	Lidar Surface Topography
LTER	NSF Long Term Ecological Research
LUCC	Land Use and Cover Change
MAST-DC	Modeling and Synthesis Thematic Data Center
MODIS	Moderate Resolution Imaging Spectroradiometer
NACP	North American Carbon Program
NASA	National Aeronautics and Space Administration
NEON	National Ecological Observatory Network
NGO	Nongovernmental Organization
NIST	National Institute for Standards and Technology
NOAA	National Oceanic and Atmospheric Administration

NPOESS	National Polar-orbiting Operational Environmental Satellite System
NPP	NPOESS Preparatory Project
NRC	National Research Council
NSF	National Science Foundation
OCB	Ocean Carbon and Biogeochemistry
OCB-SSG	Ocean Carbon and Biogeochemistry Scientific Steering Group
OCCC	Ocean Carbon and Climate Change
OCCC-SSG	Ocean Carbon and Climate Change Scientific Steering Group
OCO	Orbiting Carbon Observatory
OCO-2	Orbiting Carbon Observatory 2
OOI	NSF Ocean Observatories Initiative
ORNL	Oak Ridge National Laboratory
PACE	Pre-ACE (Aerosol-Cloud-Ecosystems)
PALSAR	Phased Array type L-band Synthetic Aperture Radar
RCP	Representative Concentration Pathway
REDD	Reducing Emissions from Deforestation and Degradation
SCIAMACHY	Scanning Imaging Absorption Spectrometer for Atmospheric Cartography
SeaWiFS	Sea-viewing Wide Field-of-view Sensor
SMAP	Soil Moisture Active and Passive
SOCCR	State of the Carbon Cycle Report
SOLAS	Surface Ocean Lower Atmosphere Study
TCCON	Total Column Carbon Observing Network
TES	Tropospheric Emission Spectrometer
UGEC	Urban and Global Environmental Change
USDA	U.S. Department of Agriculture
USGCRP	U.S. Global Change Research Program
USGS	U.S. Geological Survey
VIIRS	Visible Infrared Imager Radiometer Suite
VOS	Volunteer Observing Ships
WOCE/JGOFS	World Ocean Circulation Experiment/Joint Global Ocean Flux Study

Appendix E:
Acknowledgements

This document represents the diverse contributions of a large number of people. This effort was led by a working group made up of 25 members, of whom all have contributed ideas and many have contributed text. Nonetheless, the four co-chairs have tried to organize, synthesize, summarize, and integrate all contributions, and they accept primary responsibility for the final tone and content of this document.

The intent from the beginning has been to produce a research agenda that truly represents the scientific perspective of the active research community and we are grateful for the number of committee members from all disciplines who have contributed generously of their time and ideas. Through many discussions, presentations and other interactions with the scientific community at large, and through the public comment process, we have received a great deal of helpful input, including valuable discussions that helped to shape the Plan and determine the breadth and balance contained within. International colleagues were particularly important in helping us think about how U.S. science fits into a global research picutre. We are grateful for the wisdom and concerns shared by so many of our friends and colleagues.

It is impossible to adequately acknowledge the magnitude of the contributions from committee members (listed in Appendix B) or from those colleagues who are represented in this document only by their ideas, but there are four additional groups/individuals to whom we owe a special debt. In addition to their financial support, the members of the Carbon Cycle Interagency Working Group have been remarkably generous with their ideas, insights, and patience and we are grateful; Dennis Hansell and the members of the U.S. Carbon Cycle Science Steering Committee have been similarly helpful with their direction and input; our commitments of time and energy to this effort could not have gone forward without the support of our home institutions; Finally, Roger Hanson (Director of the U.S. Carbon Cycle Science Program office) has been with us every step of the way with support and wisdom on everything from history, context, intellectual content, and hotel and restaurant arrangements. Thanks to all!

www.ingramcontent.com/pod-product-compliance
Lightning Source LLC
Chambersburg PA
CBHW080644180526
45168CB00008B/3299